B.D. BARNES
BAKER BARNES ASSOC., INC.
9/04

"Within the corporate enterprise today, efforts to optimize operations and improve productivity through the effective deployment of IT have occurred in most operational divisions, but one area that has yet to receive such attention is R&D operations. In his book *Next Generation Product Development*, Michael McGrath offers the reader convincing evidence that this is about to change."

John Moore
Vice President and General Manager, Enterprise Solutions
ARC Advisory Group

"Powerful, insightful, and fascinating. It seems intuitive, but the vision of a seamlessly integrated NPD environment has eluded many. McGrath is among the first to describe a new level of NPD effectiveness enabled by the integration of idea, process, portfolio, resources, and financials management through recent advances in information technology."

Al Choperena
Vice President of R&D, Advanced Sterilization Products
Johnson & Johnson

"The analysis of stages in the areas of resource, project, and portfolio management was very useful in seeing how the product development process is evolving. This book provides insight about how you might move from one stage to another."

Mark Kleinle
Director of Process Improvement
Shure, Inc.

"McGrath's observations are based on his personal experience in driving the previous generation of innovation in product development. In this book he provides practical advice on how to create competitive advantage through better management of the R&D process. If his prior history is any guide, we will be using his language and tools to improve R&D productivity for years to come."

Mark Bregman
Executive Vice President, Product Operations
VERITAS Software Corporation

"In the current market, where innovation and time to market are of the essence, Mike McGrath's book provides us with thought-provoking ideas on how to manage product development moving forward."

Craig Flower
Vice President & GIO, Global Operations
Hewlett-Packard Company

"This book is poised to become a new prerequisite for corporate leadership in product development. Most of the techniques explored in it are detailed enough to provide a practical blueprint for both implementers and vendors of PLM technology."

Vasco Drecun
Research Program Director
D. H. Brown Associates, Inc.

"Michael McGrath provides an effective stage model for understanding the progression of maturity of the resource management process. He gives us a starting point, where we can get bang for the buck early, without the heavy burden."

Kirk Hemingway
Senior Manager
Avaya Inc.

NEXT GENERATION PRODUCT DEVELOPMENT

How to Increase Productivity, Cut Costs, and Reduce Cycle Times

MICHAEL E. McGRATH

McGraw-Hill

New York   Chicago   San Francisco   Lisbon   London
Madrid   Mexico City   Milan   New Delhi   San Juan
Seoul   Singapore   Sydney   Toronto

This book is printed on recycled, acid-free paper containing a minimum of 50% recycled
de-inked fiber.

To my family:
My wife, Diane, and
wonderful young daughter, Molly;
My grown children—Mike, Chris, and
Jill, and her husband, Carmen;
And my grandchildren, Matty and Callie

C O N T E N T S

Preface   xvii

SECTION ONE

INTRODUCTION

Chapter 1

Previous Generations and the Promise of the New   3

Invention and Commercialization Generation   5
Project Success Generation   9
 Project Scheduling   9
 Project Team Organization   10
Time-to-Market Generation   12
 Phase-Based Decision Making   12
 Cross-Functional Teams   14
 Standard Development Processes   15
 Portfolio and Pipeline Management   16
 Benefits of the TTM Generation   17
 Adoption of TTM Generation Practices   18
Genesis of the R&D Productivity Generation   20

Chapter 2

The R&D Productivity Generation   23

Information Transforms Strategy, Processes, and Practices   24
 The MRP Example   24
 The Integrated Accounting Example   26
 The Operation Iraqi Freedom Example   27
Characteristics of the R&D Productivity Generation   28
 Product Strategy and Portfolio Management   29
 Resource Management   30
 Project Management   32

Benefit Model of the R&D Productivity Generation   33
 Improved R&D Capacity Utilization   35
 Increased Project Resource Efficiency   35
 Developer Productivity   36
 Gains from Partnering/Outsourcing   37
 Increased R&D Administration Efficiency   38
 Pipeline Effectiveness   38
 Improved Product Strategy and Product Success   38
 Resulting Benefits   39
Development Chain Management Systems   41
Common Concepts   43
 Management Process Levels   43
 Stages of Maturity   45
Commercial Robotics Incorporated   45

SECTION TWO

RESOURCE MANAGEMENT

Chapter 3

**Improving R&D Productivity through Increased
Capacity Utilization   49**

Measuring the Opportunity to Increase Capacity Utilization   51
The New Paradigm for R&D: Capacity Management   54
Understanding R&D Utilization   55
Measuring Trends and Goal Setting   57
 Trend Reporting and Goal Setting   57
 Basis for Computing Utilization: Assignments versus
  Actual-Time Utilization   59
 Levels of Reporting   60
Best Practices for Improving Utilization   61
 Capacity-Based Portfolio and Pipeline Management   62
 Increased Awareness of Capacity Utilization   63
 Resource-Based Phase-Review Decisions   63
 Resource-Based Project Planning   64
 Resource Group Assignment Balancing   65
 Longer-Term Capacity Planning and Management   65
 Skill Planning   66
 Resource Charge-Out   66
 R&D Outsourcing   67
Summary   67

Chapter 4

Stages of Maturity in Resource Management   69

Three Levels of Resource Management   70
  Level 1: Project Resource Scheduling   71
  Level 2: Resource Requirements Planning and Management   72
  Level 3: Workload Management   73
  Flow between Levels   74
Resource Management Stages of Maturity   74
  Stage 0: Informal Resource Management   76
  Stage 1: Short-Term Utilization Management   77
  Stage 2: Medium-Term Capacity Planning and Management   78
  Stage 3: Resource Requirements Planning (RRP)   79
  Stage 4: Integrated Resource Management   79
  How the Levels and Stages Work Together   80
Roadmap for Improving Resource Management   81
Summary   82

Chapter 5

Stage 1—Short-Term Utilization Management   83

Resource Assignment   84
  Project Resource Assignment   85
  Resource Group Assignment Management   87
  Utilization Reporting   91
Resource Assignment Costs   92
Requirements   93
Benefits   96
Summary   97

Chapter 6

Stage 2—Medium-Term Resource Capacity Planning and Management   99

Resource Capacity Planning   100
  Short-Term Resource Capacity Planning   102
  Medium-Term Resource Capacity Planning   103
  Integration with Annual Financial Planning   104
  Pipeline Capacity Management   105
  Long-Term Resource Capacity Planning   106
Project Resource Needs   106
  Project Planning with Resource Needs   107
  Resource Needs and Skill Categories   109
  Resource Group Management with Resource Needs   110

Resource Transaction Process   112
    Resource Needs   113
    Resource Requests   113
    Resource Assignments   115
Requirements   115
Benefits   117
Summary   118

Chapter 7

Stage 3—Project Resource Requirements Planning and Management   119

Project Resource Requirements Planning (RRP)   122
    Using Resource Requirement Guidelines   123
    Making Preliminary Resource Estimates   123
    Time-Phasing Steps to Balance Resources   126
    Balancing Resource Requirements among Skills   128
Transforming Resource Requirements into Resource Needs   129
Resource Requirements Management   131
Workload Reconciliation   134
Requirements   136
Benefits   137
Summary   138

Chapter 8

Stage 4—Fully Integrated Resource Management   139

Level 1 Points of Integration   141
    Integration with Pipeline/Portfolio Management   141
    Reconciliation with Functional Budgets   142
    Integration with Annual Financial Planning   142
    Integration with Project Budgets   143
    Product Strategy Integration   143
    Integration with HR Systems Skill Categories   144
    Integration with Multiple Resource Groups   144
    Integration with External Resources   144
Level 2 Points of Integration   146
    Integration with Project Planning   146
    Closed-Loop Time Collection   146
    Integration with Knowledge Management Systems   148

Level 3 Points of Integration   148
Requirements   149
Benefits   149
Summary   149

SECTION THREE

PROJECT MANAGEMENT

Chapter 9

Stages of Project Management   153

Stage 3—Enterprise Project Management   155
    Enterprise Project Planning and Control   155
    Networked Project Teams   156
    Enhanced Phase-Review Process   157
Stage 4—Advanced Project Management Practices   158
    Integrated Financial Planning and Project Budgeting   158
    Distributed Program Management   158
    Collaborative Development Management   159
    Context-Based Knowledge Management   159
Summary   160

Chapter 10

Enterprise Project Planning and Control Process   161

Enterprise Project Planning Architecture   163
    Enterprise System versus Project Planning Tool   164
    Hierarchical Structure   164
    Top Down versus Bottom Up   166
    Distinction between Project Planning and Work Management   167
    Integration with Resource Management   168
The Enterprise Project Planning and Control Process   169
    Collaborative Project Planning   170
    Collaborative Project Control   172
    Integration with Project Planning Standards   173
    Integration of Project Plans   174
    Project Plan as the Basis for the Networked Team   175
Enterprise Project Planning and Control at CRI   175
Benefits   178
Summary   179

Chapter 11

Networked Project Teams   181

Cross-Functional Core Team Management   182
 Traditional Cross-Functional Core Team Model   183
 Distributed Email Communications in Core Teams   185
The Networked Project Team   188
Project Communication in Networked Project Teams   189
Project Coordination in Networked Teams   191
 Project Schedule-Items   192
 Project Calendar-Items   193
 Project Deliverables   194
 Project Action-Items   194
 Issue-Resolution Items   195
 Project Bulletin Boards   196
Content Management in Networked Project Teams   197
 Project Document Management   197
 Collaborative Document Preparation   198
Networks of Networked Teams   199
Requirements   200
Benefits   201
Summary   202

Chapter 12

Enhanced Phase-Review Management   203

Phase Reviews   203
 Product Approval Committee (PAC)   204
 Phase-Review Process   204
 Phase-Review Process Limitations   206
Enhanced Phase-Review Management Improvements   207
 Phase-Review Decisions with Resource Availability Information   207
 Combined Phase Reviews   209
 Project Tolerances   209
 PAC Information System   211
 Focus of Phase Review   212
Requirements   213
Benefits   213
Summary   214

Chapter 13

**Integrated Financial Planning and Project Budgeting
for New Products   215**

Integration of Financial Information   217
    Common Planning Models   220
    Project Budgets Automatically Derived from Resource Assignments   220
    Financial Data Directly Based on Source Data   221
    Incorporation of Financial Information in Project Documents   222
    Project Budget Updates with Actual Costs   222
    Consolidation of Financial Information across Projects   223
Integrated Financial Planning at CRI   224
    Revenue Projection   224
    Product Cost Estimate   227
    New Product Financial Plan   229
Integrated Project Budgeting at CRI   230
    Development Resource Cost   230
    Expense Budget   231
    Project Budget   231
    Capital Budget   233
    Project Financial Analysis   233
Financial Management of New Product Project   236
    Actual-Cost Management   236
    Earned-Value Analysis   238
    Revision Control   239
Requirements   240
Benefits   241
Summary   241

Chapter 14

Distributed Program Management   243

Product Development Programs   245
    Coordinating Multiple Versions of a Product   246
    Managing a Complex Project as Subprojects   247
    *Integrating Common Technology Development Projects with
        Multiple Product Development Projects   247*
    Coordinating Related Product and Process Development   248
Distributed Program Planning and Control   249
Distributed Program Team Coordination   250
Integrated Program Financial Management   253

Requirements   256
Benefits   257
Summary   257

Chapter 15

Collaborative Development Management   259

Categories of Collaborative Development   260
    Customers   261
    Development Partners   264
    Contractors   265
    Suppliers   266
    Channel Partners   267
Collaboration Systems Services   267
    Collaboration Based on Email   268
    Using a Shared File Server   269
    Collaborating with a Web-Based Workspace   269
    Integrating Partnering Capabilities into the DCM System   270
    Integrating DCM systems   272
Outsourcing of R&D   274
Requirements   274
Benefits   275
Summary   276

Chapter 16

Context-Based Knowledge Management   277

Types of Knowledge and Experience in R&D   278
    Standards, Guidelines, and Policies   279
    Templates   281
    Financial Models   282
    Experience Gained from Previous Projects   282
    Reference Documents   283
    Project Plans and Step Planning   284
    Educational Materials   284
    External Information   285
Delivering Product Development Knowledge through DCM Systems   286
    Step Libraries   286
    Linking Knowledge to Tasks and Roles   287
Collecting Product Development Knowledge   289
Requirements   290
Benefits   290
Summary   291

SECTION FOUR

PORTFOLIO MANAGEMENT AND PRODUCT STRATEGY

Chapter 17

Stages of Portfolio Management and Product Strategy   295

Process Levels of Portfolio Management and Product Strategy   297
    Product Strategy Process Level   298
    Portfolio and Financial Management Process Level   299
    Pipeline Management Process Level   300
    Integration with Other Processes   301
Stages of Portfolio Management and Product Strategy   302
    Stage 0—No Portfolio or Pipeline Management   303
    Stage 1—Periodic Portfolio and Pipeline Management   303
    Stage 2—Dynamic Portfolio and Pipeline Management   305
    Stage 3—Comprehensive Financial Management of R&D   306
    Stage 4—Integrated Product Strategy   307
Summary   308

Chapter 18

Dynamic Portfolio and Pipeline Management   309

Dynamic Portfolio Management   311
    Real-Time Integration with Project Data   311
    On-Demand Portfolio Analysis   313
    Integrated Portfolio Management Process   314
Dynamic Pipeline Management   315
    Real-Time Integration with Project Resource Management   316
    Pipeline Simulation and Optimization   318
    Integrated Pipeline Management Process   320
Requirements   321
Benefits   321
Summary   322

Chapter 19

Comprehensive Financial Management of R&D   323

The CFO's New Role in R&D   324
Consolidation of New Product Financial Information   326
    Consolidated Revenue Projection   327
    Consolidated Capital Plan   330
    Consolidated New Product Profitability and Expense   331
    Consolidated Project Budgets   331

Reconciliation of Functional and Project Budgets   331
 Tracking Functional Budget Allocation to Projects   332
 Basing Functional Budgets on Expected Resource Requirements   334
Requirements   335
Benefits   335
Summary   336

Chapter 20

Integrated Product Strategy Process   337

A System and a Process for Product Strategy   338
Product Line Plan   340
 Planned Products   341
 Standard Planned Product Profiles   343
 Simulating the Feasibility of a Product Line Plan   344
Idea Management   344
Technology Planning   347
Requirements   348
Benefits   349
Summary   349

SECTION FIVE

CONCLUSION

Chapter 21

Getting Started   353

DCM Systems   353
Processes and Practices   354
Implementation Planning   355
Conclusion   357

Endnotes   359
Glossary   361
Index   365

PREFACE

Back in 1987, I realized that there was an exciting opportunity to improve product development management. All companies were frustrated with the general unpredictability of product development and the length of time it took to translate product opportunities into product plans, and those plans into market-ready products. The solution to these frustrations lay in an extraordinarily obvious concept: Product development is a process, and, like any management process, it can be improved to achieve better results. Up until that time, product development was thought to be an art rather than a process, as remarkable as that now seems. Almost universally, and almost instantly, product development executives recognized the truth of this simple new concept: Product development is a manageable and improvable process.

Behind this simple concept were some complex details, however. The product development process needed to be defined and implemented. Development project steps needed to be codified in a logical process framework. Projects needed to be managed within a common decision-making process. And a new organizational model was needed for project teams and for senior management.

In response to these needs, in the mid-1980s, I created the basic concepts and frameworks of the PACE® (Product And Cycle-time Excellence®) process, and Pittiglio Rabin Todd & McGrath (PRTM) went on to expand, refine, and implement PACE at hundreds of companies. Little did I realize at the time that this was the beginning of a new generation of product development. PACE, and similar product development process architectures, changed the way product development was done, and over the next 15 years almost every company adopted a product development process.

By the late 1990s, I began to wonder what was next. Was there another generation of product development management on the horizon, or was it just a matter of making incremental improvements from then on? Here again, the answer was extraordinarily obvious: Information technology could completely change the way product development was managed. It had transformed every other business process, and it could do the same for product development.

This realization led to a five-year journey to define this next generation of product development. In 1998, I founded a software company, Integrated Development Enterprise (IDe), to design and develop what I refer to as a development chain management system. In addition, I worked with PRTM consultants to define some of the new management practices needed in the next generation. Finally, I had the opportunity to capture the thoughts of some of IDe's customers and PRTM's clients on the next generation. This book is the result of these five years of experience and development.

Like the previous generation, this next generation requires many new concepts and frameworks. One striking difference is how much more complex this generation is from the previous one. Management is sometimes a science. It's not always about simple practices that are easy to understand and apply. New management practices can in some cases be difficult and complex, and this is one of those cases. This book is difficult, for which I make no excuses or apologies. The next generation of product development management is many times more complex than its predecessor. It is built upon some new concepts and frameworks. It entails the introduction of many new management practices, and a lot of new terminology is needed to define these practices. Finally, the next generation is built upon an underlying information system for product development, and understanding this system requires mental effort as well.

The good thing about this complexity is that those who master it early will achieve competitive advantages that will not be easy to copy. Companies that did not embrace the practices of the previous generation of product development eventually lost out to those that did, and I expect that will be the case in this next generation as well. So it's one of those opportunities in business where great companies see an opportunity, master it before others, and achieve a sustainable competitive advantage.

This is a book about the future, not the past, and no book about the future can be as accurate as one that documents the past. I fully expect that over the next decade some of the practices described in this book will evolve differently than I anticipated. I also expect that others will evolve to levels that I had not even considered. Perhaps I'll incorporate some of this evolution in future editions. For now, however, I suggest that you take from this book the lessons that you think will work in your company, and discard the others.

This next generation of product development is only starting, but a number of companies are already off and running. As I mentioned, through PRTM, and through Integrated Development Enterprise, I've had the opportunity to learn from the early experiences of several dozen companies. Since it's still too early to describe these experiences as comprehensive case studies complete with results, and because many of these companies understandably want to keep their experiences with this next generation of product development to themselves, I don't use any of these experiences as direct examples. Instead, I use a hypothetical company throughout the book to illustrate the new practices as I describe them, so the reader has the advantage of continuity.

The book is divided into five sections. The first describes previous generations of product development management and introduces this new generation, which I call the R&D Productivity Generation. The next three sections describe the substance of this new generation, divided into three broad areas: resource management, project management, and portfolio management and product strategy. The final section is a brief chapter on getting started.

As always, many talented people helped me with this project. Michael Lecky edited the entire manuscript and made my sometimes incomprehensible words much more readable. Catherine Dassopoulos, my editor at McGraw-Hill on some of my previous books, supported me again with this project. Mark Deck and Mike Anthony, two of my partners at PRTM, reviewed many of the chapters and provided their wisdom and extensive experience. Victoria Cooper, PRTM's Director of Corporate Communications, lent her usual brilliance in discussions of content and packaging. PRTM's skilled graphics staff, including Richard Aguilar and Meredith Kay, applied their talents to the charts and figures. My Executive Assistant, Beth Reed, helped to make the many revisions and keep me from losing control of the book as it was edited and revised.

Over the last five years, I've had the privilege of working closely with Richard Moore, IDe's CEO, to create and bring to market a comprehensive system for development chain management. His extensive experience in product development and his leadership were essential in turning this vision into a reality. Several people at IDe helped to brainstorm some of the new concepts contained in the book. IDe's extensive customer base and its experience with this new generation of development were invaluable. In particular, I want to express my appreciation to Joanna Ladakos and Ralph Brown.

As with my other books, I owe a lot to my wife, Diane, and to my youngest daughter, Molly, who gave up much of one summer with me while I worked on the manuscript.

Michael E. McGrath

Introduction

1

C H A P T E R

Previous Generations and the Promise of the New

In the 1990s, time-based strategies changed the competitive balance in many industries as some competitors radically improved time to market through new management practices. These practices marked an entirely new generation of product development management, and eventually all remaining companies adopted these practices in order to survive. We are now entering another generation of product development management. This time, the focus is on productivity—developing more new products than your competitors and doing it with a lower investment. I call it the R&D Productivity Generation, and it will launch new competitive strategies.

The benefits of this new generation come at just the right time. The decline in the economy at the turn of the century (2000) refocused management's attention on increased productivity, and the opportunities for R&D productivity just may be on the top of this list. The competitive advantages of dramatic increases in R&D productivity will have the same strategic impact that dramatic increases in time to market had in the 1990s.

While this book is about the future of product development, somehow I think it seems appropriate to start a book about the future by looking backwards. Product development has changed a lot over the last century, and a brief review of its history provides an important perspective on its transformation over time and how we got to where we are today. Hopefully, it will also inspire you to embrace the changes ahead in the next generation of product development management.

Although somewhat arbitrary, it is useful to look at the history of product development management as an evolution through distinct generations, with each generation initiating improvements based on its signature principle or focus. In fact, each new generation achieved not just progress, but some form of transformative breakthrough that changed the competitive balance in favor of those enterprises that took advantage of the breakthrough before their competitors did. Over time, of course, these cutting-edge practices became standard practices, until the point was reached where the practices ceased to confer any real competitive advantage. Every time that point has been reached, product development management has evolved to its next form.

We'll look at four generations of product development management, as illustrated in Figure 1–1. It is crucial to note that the rate of change in product development management has accelerated. The first generation took more than 50 years to run its course. The second took more than 30 years. But the current generation reached broad-based adoption in just 15–20 years. I expect that the philosophy, practices, and enabling systems of the new product development generation

F I G U R E 1–1

Four Generations of Product Development Management

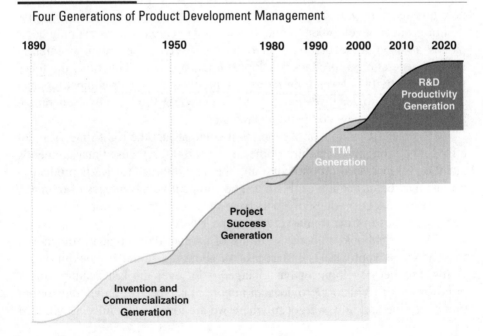

will be fully adopted in a little more than a decade, and will be a decisive competitive factor for some companies within a few years. In addition to the accelerating pace of evolution, each generation of product development management has been substantively more complex than its predecessor. And companies that are innovators of this transformation have visionary leaders who see the future of product development, understand its importance, and are willing to do the hard work to get there. While each generation achieved significant advantages, it also required a willingness to change and depart from current practices.

My classification of generations, though certainly imperfect, is accurate and serviceable enough for the purposes of our discussion. Going back in history, I refer to the first generation of product development management as the Invention and Commercialization Generation. It began toward the end of the 19th century and lasted until approximately 1950. As the name implies, the first attempts to institutionalize and commercially translate new inventions into mainstream products were made during this period. The year 1950 marked the inception of what I call the Project Success Generation, whose focus turned from

> The Time-to-Market Generation dates back to the mid-1980s; its primary focus is developing individual products faster.

management of the development laboratory to management of individual development projects. This is when rudimentary practices for project planning were first introduced, and individual project teams first organized.

The development management precepts and practices with which most development personnel are currently familiar comprise what I refer to as the Time-to-Market Generation, whose primary focus is developing individual products faster. The TTM Generation dates back to the mid-1980s, and the vast majority of the people currently involved in product development management, i.e., those with less than 20 years' experience, have the impression that this is the way R&D has always been managed. It hasn't. Today's practices are relatively new. Moreover, and as I hope to show throughout this book, they are already being replaced by the practices of this next generation, which I call the R&D Productivity Generation because of its emphasis on the creative productivity of the entire development enterprise.

INVENTION AND COMMERCIALIZATION GENERATION

Product development obviously predates the 20th century, but efforts to formally encourage it down particular paths, much less organize it, do not appear to have predated the Industrial Revolution. Perhaps the first organized attempt at what could remotely be described as product development management was made in the early 1700s, when the British government established its "Board of Longitude." The board offered a monetary prize, equal to tens of millions of dollars today, to the first developer of a practical method for determining longitude at sea. The board also established a scientific jury to assess the work of anyone claiming the prize, which was ultimately won by a clockmaker named William Harrison. It was not until the latter 19th century, however, that anyone seriously proposed that invention might be an organizable process, as opposed to a random event—an individual act of mind. The first formal generation of product development management can reasonably be attributed to Thomas Edison, and his creation of the first commercial laboratory. In 1887, Edison set up the world's first full-fledged research and development center, in West Orange, New Jersey. This laboratory was the first of its kind anywhere, and it is considered by some to be Edison's greatest invention. The new laboratory complex, consisting of five buildings, opened in November of that year, with a three-story main laboratory building containing a power plant, machine shops, stockrooms, rooms for performing various types of experiments, and a large library. Four smaller, one-story buildings housed a physics lab, chemistry lab, metallurgy lab, pattern shop, and chemical storage areas. The laboratory allowed Edison not only to work on any sort of project, but also to work on as many as 10 or even 20 projects at once. Edison said himself that the purpose of this laboratory was to commercialize inventions. "My main purpose in life is to make money so I can afford to go on creating more inventions." In other words, Edison identified commercialization as the "competency," to use the contemporary term, that would open the door to unlimited opportunities for innovation. That view, it so happens, is a deeply held value of the emerging R&D Productivity Generation, as we'll see in later chapters.

In his lifetime, Edison patented 1,093 inventions, earning him the nickname "The Wizard of Menlo Park." The most famous of his inventions was the incandescent light bulb, but he also developed the phonograph and the "kinetoscope," (an early version of motion pictures.) He improved upon the original design of the stock ticker, the telegraph, and Alexander Graham Bell's telephone. Edison also professed the notion that product development was much more than just a bright idea; that it involved a lot of work. "Genius," according to the famous quote, "is 1 percent inspiration and 99 percent perspiration." In that regard, Edison's lab in West Orange was more of a product development process than a palace to genius.

While Edison and his developers were perspiring so heavily in New Jersey, others were also laboring to create and commercialize other innovations.

On December 17, 1903, near Kitty Hawk, North Carolina, Wilbur and Orville Wright achieved the world's first flight of a powered, heavier-than-air machine. With Orville at the controls, the plane flew 120 feet in 12 seconds. The Wrights believed that they could commercialize their invention, which would eventually be used to transport passengers and mail. When they first offered their machine to the U.S. government, they were not taken seriously, but by 1908 they got a contract with the U.S. Department of War for the first military airplane. Thus began the commercialization of the airplane.

Also around the turn of the 20th century, new companies were created to capitalize on the commercialization of innovative solutions to specific problems. Although IBM was incorporated on June 15, 1911 as the Computing-Tabulating-Recording Company (C-T-R), its origins can be traced back to developments at the close of the 19th century. The first dial recorder was invented by Dr. Alexander Dey in 1888, and it became one of the building blocks of C-T-R. The Bundy Manufacturing Company was incorporated in 1889 as the first time-recording company in the world, and it later became part of C-T-R. Perhaps even more significant for IBM, a U.S. government agency requirement late in the 19th century led directly to the development of one of the company's principal lines of business. During the height of the American Industrial Revolution, when the United States was receiving waves of immigrants, the U.S. Census Bureau recognized that its traditional counting methods couldn't handle the swelling volume of census data. As a result, the bureau sponsored a contest to find a more efficient means of tabulating census data. The winner was Herman Hollerith, whose Punch Card Tabulating Machine used an electric current to sense holes in punched cards and keep a running total of the data. Capitalizing on his success, Hollerith formed the Tabulating Machine Company in 1896.

The Ford Motor Company had similarly modest beginnings. On June 16, 1903, Henry Ford and 11 business associates, armed with $28,000 in cash, gave birth to what would become one of the world's largest corporations. If the Ford Motor Company's single greatest achievement was making automobiles affordable, then the key to that achievement was the moving assembly line. First implemented at the Highland Park, Michigan plant in 1913, this process innovation allowed individual workers to stay in one place and perform the same task repeatedly on multiple vehicles as they passed by. The line proved tremendously efficient, helping the company far surpass the production levels of its competitors—and making the vehicles more affordable. Henry Ford insisted that the company's future lay in the production of affordable cars for a mass market. The production efficiencies he achieved helped usher that future into being.

Since its founding in 1925, Bell Laboratories has helped shape the ways in which people live, work, and play. Today's countless millions of transistors are all descended from the first transistors invented by Bell Labs between 1947 and 1952. Similarly, today's digital communication of sound, images, and data rests on mathematical foundations of information theory derived by Bell Labs' researchers in the late 1940s.

We also shouldn't forget the early commercialization of the products we take for granted. In early 1894, the Hershey Chocolate Company was born as a subsidiary of Milton Hershey's Lancaster, Pennsylvania caramel business. In addition to chocolate coatings, Hershey made breakfast cocoa, sweet chocolate, and baking chocolate. In 1900, he sold the Lancaster Caramel Company for $1 million, but retained the chocolate manufacturing equipment and the rights to manufacture chocolate, believing that a large market existed for affordable confections that could be mass-produced. He returned to his birthplace, Derry Church, and located his chocolate manufacturing operation in the heart of Pennsylvania's dairy country, where he could obtain the large supplies of fresh milk needed to make fine milk chocolate. In 1903 he opened the world's largest chocolate-manufacturing plant.

Companies formed with a commitment to continuing invention and commercialization prospered during the Invention and Commercialization Generation. In 1885, Edwin Binney and C. Harold Smith formed a partnership and called their company Binney & Smith. Early products included red-oxide pigment used in barn paint and carbon black for car tires. In 1900, the company began producing slate school pencils in its newly opened Easton, Pennsylvania mill. Listening to the needs of teachers, Binney & Smith introduced the first dustless school chalk two years later. In 1903, after observing a need for safe and affordable wax crayons, the company produced the first box of eight Crayola® crayons, selling it for a nickel.

The research and development organizations in most companies evolved gradually throughout the first half of the 20th century, and General Motors provides a very good example. In 1911, GM created a laboratory for materials analysis and testing. It gathered together a research staff in the early 1920s, and then an engineering staff in the 1930s. Dayton Engineering Laboratories, headed by Charles Kettering, had invented the first practical electric self-starter, in 1912. Dayton was acquired by General Motors in 1920, and became the General Motors Research Corporation. GM did not have a department or section of the corporation explicitly devoted to engineering until 1931. It formed the General Technical Committee in 1923 to unite various research and engineering organizations. In 1945, General Motors created a central engineering function, whose task was to incorporate the fruits of GM's research into new products at a faster pace. Throughout the 1940s, attention turned to the invention and production of armaments needed to fight World War II, and the focus of product development was on military programs. When attention returned to commercial products, the lessons learned in military program management inspired a new focus on project management.

PROJECT SUCCESS GENERATION

The 1950s generally marked the emergence of what I've chosen to call the Project Success Generation because companies focused on what it took to make new product projects successful. Many new product projects were bold, and success rates were not always very high, so success was the only objective, no matter what it took or how long it took. This was the point at which private enterprises, as well as major government programs, began to increase their focus on the management of individual projects within their research and development organizations. Characteristic of this focus was the development of new project-scheduling practices and the early formation of project teams.

Project Scheduling

The earlier generation of development management produced some new project-scheduling techniques. The most notable of these was the Gantt chart created by Henry Gantt, an industrial engineer with the U.S. Army during World War I. Gantt charting was used to divide a project into phases, with each phase depicted as bars or lines that showed the status of that phase as compared to the other project phases.

The late 1950s saw an increased emphasis on project scheduling, and particularly on the use of mathematics in scheduling. Two mathematical methods for project planning and scheduling became popular: the Critical Path Method (CPM) and the Program Evaluation and Review Technique (PERT). CPM was originally applied at DuPont, and was used early in the Project Success Generation to focus attention on the activities whose completion was of the highest priority. PERT, another duration-analysis technique, was developed by the U.S. Navy in conjunction with the consulting firm of Booz, Allen, and Hamilton.

CPM is a network analysis technique used to predict project duration by analyzing which sequence, or "path," of activities has the least amount of scheduling flexibility, or "float." Early dates are calculated by means of a forward pass, using a specified start date. Late dates are calculated by means of a backward pass, starting from a specified completion date, usually the forward pass's calculated finish date.[1]

PERT is an event-oriented network analysis technique that is used to estimate program duration when the individual activity's duration estimates are uncertain. PERT applies the critical path method using durations that are computed by a weighted average of optimistic, pessimistic, and most likely duration estimates. PERT computes the standard deviation of the completion date from those of the path's activity durations.[2]

CPM and PERT had some value in getting product developers to think and plan beyond the development of a project schedule, especially on large and complex projects, but the two techniques were of limited use in guiding project teams. I recall one case, in the early 1970s, when I met with the manager of a large project team. The entire team of more than 75 developers worked at desks in the same large room. On one wall of this room, visible to all, was a gigantic PERT chart, maintained by two full-time clerks who listed hundreds of tasks. Yet, despite this heroic effort to use the PERT chart to inform the developers of what they needed to do, there was a line of 20 people outside the project manager's office waiting for him to tell them what they should be doing that day. I was told that this queue was a daily routine.

CPM and PERT eventually found useful niches in scheduling projects. Gantt charts continued to have some practical application, but there was still no good solution for project planning and control.

Project Team Organization

Companies also began to experiment with a variety of organizational approaches to project management. The functional approach was the most prevalent for development projects. Under this approach, each function contributed to the product development process in a serial, or hand-off fashion, similar to that of a relay race. The development cycle started with Marketing devising the requirements for the product, which were then handed off to Engineering, which prepared specifications and began designing the product. Manufacturing then built prototypes, Sales began to sell the product, and Customer Service got involved at the end to support it. An extreme version of the functional approach was sometimes referred to as the "customer/vendor" approach, by which the efforts of the successive development functions were integrated through formal documents or even contracts. Under this approach, each functional organization acted independently, contracting out its services to the other organizations within the company. These quasi-autonomous transactions were conducted formally, with many approvals along the way. Typically, this "internal market" form of project management was implemented by a company after a major project failure.

Another variation of the functional project organization used program coordinators to manage schedules and coordinate activities across the functional projects. This approach had its roots in the defense industry. The program manager's role was to act as a coordinating mechanism across the functional projects, and the primary responsibility of the program manager was to prepare schedules of tasks and activities. These schedules sometimes incorporated the integrated input from those actually working on the project, but it was all too frequently the case that schedules were devised with little or no regard for project realities.

Some companies assigned multiple functional project managers to shore up the weaknesses in the functional approach. Multiple project teams would be formed, one from each function, with multiple project managers, one from each functional team, and these managers would try to work together to co-manage the project. Each of the project managers managed their project through the entire development cycle, and they tried to work together to resolve differences.

Still another variation was to assign a project team to each phase of the project. This approach sought to take advantage of specialized skills in each phase, and it assumed that individuals performed better if they focused on a piece of the development process rather than on the entire development cycle.

The matrix management approach to project management was the first major deviation from the functional approach. The basic philosophy underlying matrix management was to take the best from the functional organization and combine it with a temporary project focus. People within a given function were "loaned out" on a full-time or part-time basis to work on a specific project. While the matrix organization was the precursor of the cross-functional project team organizational model of the Time-to-Market Generation of the 1980s and 1990s, it generally failed because of organizational conflict and lack of a clear cross-functional process.

The most extreme approach to project team organization was to give a project team absolute autonomy. Various colorful names were used to describe this approach, such as skunk works, swat teams, and tiger teams. In this approach, the team was given a general sense of direction, provided with the necessary funding, and then left alone to do what it wanted. Autonomous teams produced some great successes and some monumental failures because they were disconnected from the company's product development process, for better or for worse. Nevertheless, valuable lessons were learned from the focus, efficiency, and motivation of the autonomous project team, which would be applied in the next generation of development management.

By the middle of the 1980s, a lot of progress had been made, but there was a feeling that for product development to be successful, project management needed to be integrated into a more formal product development process.

TIME-TO-MARKET GENERATION

The breakthrough of the TTM Generation was its recognition that product development is a management process, and, as with any management process, a better process produces better and more predictable results. This generation also recognized that this development management process extends beyond project management to include the decision process that controls the launch of new projects and the use of standards across all projects to ensure more consistent project outcomes.

Most people would acknowledge that the primary objective of this generation of product development management improvement was faster time to market, hence the name I've chosen for it. From the outset, this generation recognized that a bigger financial gain could be achieved from getting products to market faster than from reducing the cost of development. First-to-market advantage was one important consideration, but there were also others, as I'll discuss later in this book.

The TTM Generation initially focused on three management areas: the process of making decisions regarding new products on a phase-by-phase basis, the introduction of cross-functional project teams, and the use of standard processes for developing products. Toward the end of this generation, some companies also added rudimentary portfolio management as they realized that they needed to set priorities across projects as well. The PACE® process[3] created in 1988 by Pittiglio Rabin Todd & McGrath (PRTM), was a leading example of the new practices of the TTM Generation.

Phase-Based Decision Making

Product development is driven by the decision-making process that determines what products to develop and how development resources are to be assigned. Through this process, senior management leads product development, implements product strategy, and empowers project teams to develop new products. Prior to the TTM Generation, this decision making was generally ineffective because it wasn't recognized as a formal process. Late and sequential decisions wasted precious resources, indecision caused projects to drift, and lack of consensus led to frequent midstream changes to product designs and the severe over-allocation of scarce resources. Developers lost respect for executive leaders, referring to them, for example, as "decision-proof."

In the middle of the 1980s, new practices emerged to improve this critical decision-making process. Stage-Gate™,[4] developed by Robert G. Cooper, was one of the first major new practices of the TTM Generation. The project contract system developed by Steve Wheelwright and Kim Clark of Harvard, and the phase-review process developed by PRTM are examples of phase-based decision processes. While each of these processes is a little different from the others, they all have the basic objective of making project investment decisions one phase at a time.

A phase-based decision process requires the review of a project at each phase of its development, based on specific criteria for completing the current phase and preparedness to enter the next phase. In most companies, the authority for making these phase-review decisions has been clarified with the establishment of a new organizational entity, referred to in PACE terminology generically as the Product Approval Committee (PAC), but sometimes referred to by other names. This group has the specific authority and responsibility to approve and prioritize new product development investments by initiating new projects, canceling and reprioritizing projects as appropriate, ensuring that products being developed fit the company's strategy, and appropriately allocating development resources. The PAC meets at the completion of each project phase, reviews the project results and plan for the next phase, and then typically convenes in closed session to make a decision. In most companies, PACs have canceled or refocused problematic projects earlier than was formerly the case, substantially reducing the waste of development resources and better focusing products on the true needs of the market. Some companies saved as much as 10 percent of their R&D budgets through timely action to prevent wasted development.

> Phase-based decision making is one of the most important innovations of the TTM Generation. It enables companies to invest in new products progressively, creates an increased sense of urgency, establishes consistent expectations of development time, and cancels projects before they waste development resources.

Under phase-based decision making, all major product development projects are required to go through a common phase-review process, which consists of standard phases for product development, generally four to six, which define the scope and expectations for each phase. The PAC approves all projects phase by phase, giving the project team the authority to continue through the next phase of development. In essence, this creates a highly efficient two-tier management organization for product development: The PAC empowers and funds the project, and the cross-functional project core team executes it.

Stage-Gate™ processes are similar to phase-review processes, but focus more on sequential gates and less on cross-functional decision making. Phase-review processes also use standards to establish the cycle time for each phase, and project teams are empowered to complete the phase within the allotted time. Stage-gate processes generally don't use cycle-time standards to create this deadline pressure. In application, there are a range of best practices applied to phase-based decision-making processes, regardless of the name used.

Phase-based decision making is one of the most important innovations of the TTM Generation. It enables companies to invest in new products progressively, creates an increased sense of urgency, establishes consistent expectations of development time, and cancels projects before they waste development resources. Cross-functional PACs also make it feasible to have effective cross-functional project teams, because they have broad authority to empower the team.

Cross-Functional Teams

A key innovation of the TTM Generation with respect to the organizational model for project teams was the introduction of the cross-functional team. PACE's cross-functional core team model proved to be so effective that core teams replaced functional teams as the standard. As previously mentioned, the functional team was the primary model for organizing project personnel from the 1950s until the mid-1980s, because it comported with the strong command-and-control hierarchical structure of companies at the time.

The functional team structure had many vertical layers and generally lacked horizontal communications. This slowed coordination and decision making, since both first had to flow up the vertical organization and then back down. The primary defect in the functional project team model lies in its very structure. It erected barriers among functional groups in what was inherently a cross-functional process.

In the TTM Generation, the cross-functional core team organizational model and the related core team management practices enabled project teams to better synchronize execution of the project by *organizing* teams more effectively. The simplified organizational structure of the core team led to dramatic improvements in project communication, coordination, and decision

making. This organizational solution solved many communications problems. Because the core team members worked together on a daily basis, they
were able to communicate rapidly and reliably. No information was lost
among the team because of organizational filtering or distortion. It greatly
simplified communications.

The cross-functional core team project organization enabled effective coordination of activities with little wasted effort. The core team was directly
empowered and responsible for the success of the project, so core team members had the responsibility and authority for the coordination of all project activities. Team members would meet regularly to coordinate project activities
among themselves, and then each core team member would coordinate his or
her activities with the members of the extended team. This hub-and-spoke
organization of the core team worked well.

Standard Development Processes

The third major area of improvement during the TTM Generation was the
application of standard development practices, especially structured development practices, across all product development projects. During the Project
Success Generation of the 1950s, 1960s, and 1970s, product development activities varied widely from project to project, even when the activities were
aimed at closely similar ends, leading to unpredictable completion dates and
wasted time.

Some companies overreacted to this pointless variation in development
activities by trying to define their product development process in excessive
detail. They tried to control everything by defining precisely how each activity
should be carried out and exactly what the output should look like. This
micromanagement was typically document-based, with the completion of each
task controlled through the preparation and approval of a specifically defined
document. This bureaucratic approach was typified by thick notebooks of standards, which in most cases were ignored, fortunately.

Eventually, a balance was struck between too little and too much structure. The structured development process of PACE is an example of this balance. Development activities are structured in a hierarchy from phases (the
highest and broadest level), to steps, to tasks, and sometimes to detailed activities. Phases are the primary level for cross-project management, and are the
same for all projects. Steps are consistent for all projects, but steps are selected
to fit each project. This step level is the primary level for project planning and
scheduling. Tasks provide guidelines for how a step can be completed, but
they can be implemented differently by every project team.

The primary advantage of structured development is its introduction of top-down project planning based on cycle-time standards. Top-down project planning uses standard steps with standard cycle times as the basis for project planning. The cycle times used in these steps are defined based on best practices, experience within and outside of the company, and industry benchmarks. The standard cycle times codified for use in project planning are aggressive but feasible times, which all project managers are expected to use in scheduling their projects. It is then left up to them to figure out how to complete the steps within the stipulated cycle times.

When all project teams applied these standards, development times were reduced and greater consistency was achieved across all projects.

This top-down approach replaced a bottom-up approach, where individual developers estimated how long it would take them to complete specific tasks, and the cycle time for the project was determined by summarizing all of the individual estimates. The individual estimates were frequently conservative, with additional times added to provide cushions, and the accumulation of these increased cycle times. It also led to wide inconsistency across projects in terms of completion dates, since developers all used different estimates and managers built in different cushions.

I emphasize the distinction between top-down and bottom-up project planning because the former was one of the big improvements of the TTM Generation, and because that improvement will be jeopardized in the next generation of product development management if not explicitly recognized and diligently preserved. We'll discuss this issue at some length later in the book.

Portfolio and Pipeline Management

Toward the end of the TTM Generation, some companies implemented periodic portfolio and pipeline management practices. After achieving increased predictability of product development projects, they began to recognize that they couldn't do everything they wanted to do and had to set priorities. They also recognized that they needed to better link their portfolio of projects with their business strategy.

With portfolio management they improved the way they selected and prioritized a group of projects to achieve business goals, especially long-term product investments. With pipeline management they prioritized project investments to better allocate resources to the best projects.

I refer to this as "periodic" because it was done periodically, usually when there was a crisis or frustration of some sort, and also relied on manually collecting information across all projects, which was time-consuming and limited. We'll discuss the limitations of portfolio and pipeline management practices of the TTM Generation in Chapter 18, but, despite these limitations, these practices were successful. Companies made better decisions across their portfolios of products and improved the value of their R&D investment as a result.

At this time also, some companies extended portfolio management even further into product strategy. They introduced platform management practices that improved the strategic focus on platforms of products, and not just individual products, and increased the leverage of common technology across all products. While in the TTM Generation these practices were somewhat limited because they were information-intensive, they provided a base that the next generation could build on.

Benefits of the TTM Generation

The gains in development productivity over the last two decades or so are quite impressive. On average, most companies reduced their time to market by 40 to 60 percent across all new products. In some industries, such as pharmaceuticals, development cycles were reduced by less than that, but were still reduced by two to three years. Consider, however, that two or three years, in the context of a blockbuster drug's period of exclusivity, can easily translate to hundreds of millions of dollars in additional earnings before the patent expires. Across all industries, this improvement was so significant that a flood of new products were introduced in the mid-1990s. Some believe that this outpouring of innovation was a contributing factor to the bubble economy of the late 1990s, and that the bubble eventually burst because, at least in part, the market could not absorb all the new products.

> The gains in development productivity over the last two decades are quite impressive. On average, most companies reduced their time to market by 40 to 60 percent across all new products.

Decreased time to market also reduced the costs of many projects. Frequently, a reduction in development time also resulted in shorter resource assignments. For example, a project completed in 12 months instead of 24 months would cost less, depending on the extent of additional resources required for a shorter time. Reducing the waste of development resources was another major benefit for most companies. Wasted development is the investment in canceled projects that could have been canceled much earlier.

The practices of the TTM Generation also improved the success of new products because the products were better targeted at customer needs. By integrating all functions in the development process from the beginning, customer requirements were more clearly understood and incorporated into new products.

The final benefit of time-based product development is the strategic advantage. It enabled time-based competition. Companies with time-to-market advantages in their product development process were able to get products to market much faster and seize market share away from competitors. They were able to respond faster to technical advances and market opportunities. In some cases, time-based competition altered the entire competitive balance of industries.

Perhaps the best quantitative measure of the improvements in development management during this generation was the change in the R&D Effectiveness Index. This metric, developed by PRTM in 1992, measures the profit from new products against the investment in new products. It's computed by taking the percentage of new products in a company's revenues (40 percent, for example), multiplying that percentage by the profit percentage (10 percent, for example), and dividing the result by the percentage of revenue invested in R&D (5 percent, for example). In order to normalize the index so that 1.0 represents break-even, i.e., $1 returned for every $1 invested, the R&D percentage is added back to the new product profit, so the R&D Effectiveness Index in this example would be 1.2 [40% × (10% + 5%) / 5%]. This roughly means that $1.20 was returned for every $1 invested, or that the return on investment was 20 percent.[5]

PRTM benchmarked the R&D Effectiveness Index six times over eight years, and computed an overall index across eight technology industries, on a weighted-average basis. The overall RDEI doubled from 0.5 in 1992 to 1.0 in 2000. The implications are quite profound. This one statistic indicates that of the approximately $200 billion investment in R&D each year, an additional $100 billion of profit per year was generated. The increase in the index also implies that the output of new products per dollar of investment doubled between 1992 and 2000, although some of the gain in the index should be attributed to more commercially successful products.

Adoption of TTM Generation Practices

As is the case with the adoption of all new management processes, practices, and systems, the innovations of the Time-to-Market Generation proliferated according to a fairly predictable adoption curve. Early innovators who want to achieve the first advantages of new practices are the first to adopt them. These early adopters are later followed by others who see the early success of the leaders and move quickly to implement the new practices to remain competitive. This large-scale adoption movement marks an inflection point in the adoption curve, as more and more companies embrace the new practices until they become standard. Toward the end of the adoption curve, there are often a few laggard companies—late adopters—that struggle to catch up before their businesses are rendered noncompetitive by the wave of change.

The adoption curve for TTM Generation principles and practices can be approximated from data collected by PRTM from a sample of more than 600 companies. As you can see from Figure 1–2, the curve approximates a shallow "S" shape, with slow but persistently increasing adoption over the first five to

F I G U R E 1–2

TTM Generation Adoption Curve

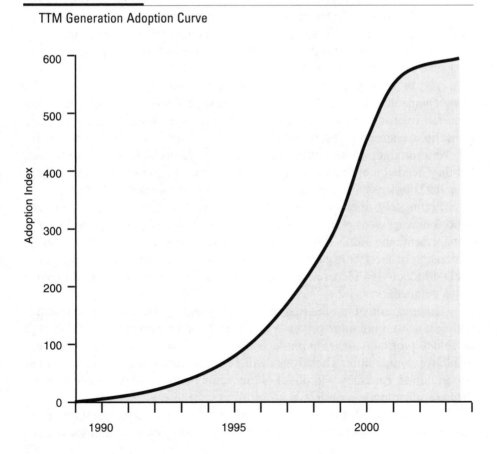

seven years, rapid adoption through the mid-1990s as more companies came to understand the competitive advantages, and a slowing of adoption over the last few years, as TTM practices became nearly universal.

By the beginning of the 2000s, the TTM Generation was approaching full adoption. Most companies had implemented these practices, and those that had not were generally in industries where R&D investment was less critical to success, such as commodity manufacturing. For the most part, companies in R&D-critical industries that were slow to adopt these practices became non-competitive and were acquired by faster competitors.

GENESIS OF THE R&D PRODUCTIVITY GENERATION

The maturing of the TTM Generation coincided with an historic downturn in the economy. For many companies, the steady rise in revenue that began to be taken for granted by the late 1990s was replaced by revenue declines at the turn of this century. Technology-intensive companies were hit particularly hard. Instead of numerous new product opportunities, which were limited only by the ability to find enough developers, new product development was constrained by tight R&D budgets.

Yet, at the same time that they were constraining R&D investments, many companies desperately needed more new products. They also found that expected returns from new product investments were lower because revenue projections were much more conservative. For many, the dilemma was painful: New product opportunities were diminished, investment was restricted, but they needed more new products. Increasingly, companies began looking at their R&D process and asking the question: How do we get more for less?

Fortunately, there is a promising answer to that question. It's the advanced management practices of the next generation of product development management, the R&D Productivity Generation. Just as the Time-to-Market Generation of the 1990s gave rise to strategies for time-based competition, the R&D Productivity Generation will give rise to development-productivity-based competition.

Companies that are the first to take advantage of these new opportunities will be able to bring more products to market than their competitors. They will be able to profitably develop products that are too expensive or not financially viable for competitors. Their R&D will be more highly responsive to market opportunities. And they will do all of this while being more profitable because they are spending less on R&D.

To some extent, the seeds of the advances in this next generation of development management were sown from the productivity limitations and compromises of the previous generation. Moving particular products from concept to market faster meant concentrating critical resources on those projects, sometimes to the extent that fewer products were developed. Efficiency was not important; speed was. Attention was taken away from productivity in order to improve time to market.

The next generation of product development management builds upon the time-based gains of the TTM Generation and will take product development to the next plane of performance in terms of the pace, volume, and quality of new product introductions. What I've chosen to call the R&D Productivity Generation of product development management has been made possible by recent advances in information technology, specifically what I will refer to broadly and generically as development chain management (DCM) systems. The fact that the next generation of product development management is driven by information technology will not be a surprise to anyone who has studied the progress of management over the last 40 years. Information systems have transformed all other areas of management, from manufacturing to accounting to customer service, and it's happening now in product development.

> The R&D Productivity Generation has been made possible by recent advances in information technology, specifically what I will refer to broadly and generically as development chain management (DCM) systems.

Again, the new focus is on productivity: a greater output of products for the same or even lower level of investment—getting more for less. Further improvements in development speed, and new gains in product quality, will be the byproducts of this productivity gain, for reasons I'll explain later on.

Figure 1–3 illustrates the potential improvement from this next generation of product development, using the R&D Effectiveness Index as the measure of product development performance. It shows the improvement achieved by the previous generation of product development management practices, as reflected in the R&D Effectiveness Index gains from 1992 to 2000, based on benchmarking done by PRTM's benchmarking subsidiary, The Performance Measurement Group. You can see how the index increased consistently as more companies adopted the new, time-based product development practices. The improvement rate across all industries was approximately 14 percent per year until 1996, when it slowed to 4 percent, after most companies had implemented the practices and achieved the benefits. By 2000, R&D Effectiveness plateaued at almost 1.0, double what it was in 1992.

F I G U R E   1–3

Product Development Process Improvement Curve

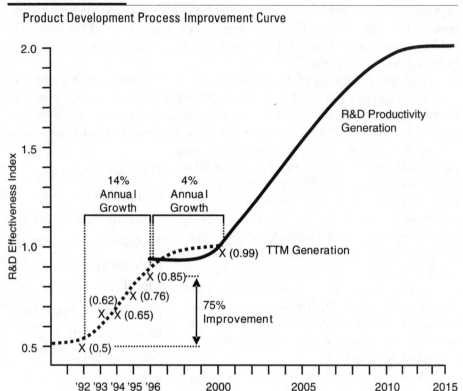

I expect that R&D Effectiveness will double again in the new generation, the R&D Productivity Generation, as is indicated by the extrapolated
process improvement curve. For individual companies, this means getting
twice as much new product profit from the same R&D investment, or the
same profit with half the investment, or some combination of the two.
Across all R&D worldwide, this roughly means that an additional $200 bil-
lion of new product profit could be achieved with the same investment level
of $200 billion per year, or that only half of the $200 billion invested today
would be needed in the future to achieve the same profit, or most likely some
combination of these.

The next chapter will offer an overview of the next generation of product
development, and the specific benefits that will enable a doubling in R&D Effectiveness. The remainder of the book will describe the practices of this new
generation in detail.

2
CHAPTER

The R&D Productivity Generation

In the new product development generation, the R&D Productivity Generation, the product development management process will be transformed into a highly efficient, highly automated, and highly integrated enterprise-wide process that will operate at extraordinary performance levels. The primary performance improvement is increased productivity, but there will be many others as well. As was the case with the previous generation of product development, this transformation is so significant that it has the potential to change the competitive balance within an industry. Very few technology-based companies that did not make the transition to the previous generation of product development are still in business today, and those who led the transition gained notable competitive advantages. Similarly, companies that make this transformation to the new practices and supporting systems before their rivals will create significant competitive advantages for themselves, and all companies will eventually need to make the same transformation if they are to remain competitively viable.

This next major advancement in product development management is much more complex than the previous one, as has historically been the case with generational improvements. The next generation goes beyond the primary focus on project management of the previous generation to include comprehensive resource management and integrated portfolio and product strategy, and will introduce many more management practices than the previous generation. The next generation of productivity-based gains requires integrated systems, which I call development chain management systems, to enable these new practices.

INFORMATION TRANSFORMS STRATEGY, PROCESSES, AND PRACTICES

Information can transform the way we do almost everything. It improves business practices, enables more productive management processes, and changes the very strategies on which businesses are based. Process improvement based on new information systems has been a constant theme for the last four decades, and it is also the underlying theme of the next generation of product development. In fact, product development, perhaps more than other business processes, is about information.

> Process improvement based on new information systems has been a constant theme for the last four decades, and it is also the underlying theme of the new generation of product development.

Let's look at some of the many times in history that information was used to transform a process or strategy. It will help explain why the new flows of information made possible by advanced systems will enable massive improvements in product development productivity.

The MRP Example

The story of materials requirements planning (MRP) provides a good historical example of how new information systems completely transformed a critical process in all manufacturing companies. Prior to the 1970s, inventory was planned and managed using techniques such as inventory re-order point, which determined the expected material and component requirements based on average historical usage data, which was the only relevant information available. With these techniques, inventory levels could be adjusted a little, but a significant improvement was not possible. Inventory levels remained very high. Material shortages delayed production, which, in turn, made revenue unpredictable. And many companies wrote off excess inventory whenever they changed their products.

The stockroom maintained records—sometimes manual and sometimes automated—of what was in inventory. Purchasing maintained records of what was on order, usually through copies of purchase orders. Production followed its own schedules. Engineering maintained bills of material. Sales maintained a sales forecast of what it expected to sell. Everyone had his or her own information, in a form that was useful, but the information could not be integrated.

Then, in the 1970s, materials requirements planning processes were introduced. These processes, based on the integrated information provided by MRP systems, revolutionized the way manufacturing was managed. Figure 2–1 illustrates the integration of information with MRP that was used to determine the volume of material and components a company should purchase. Instead of looking at the historical requirements for a component, a company was able to take its sales forecast, adjust it to determine what was called the master production schedule for the end products it needed, "explode" the product requirements into subassemblies, and net these against what was in production to get the materials requirements. It then netted these materials

F I G U R E 2–1

Materials Requirements Planning

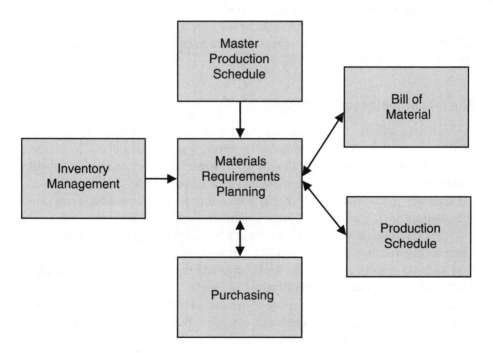

requirements against what was on hand and already on order, thereby determining what it needed to order, and when.

Beyond implementing MRP systems, companies needed to completely transform their materials management process and related processes. This transformation was a complex and difficult undertaking, and some companies did not successfully make it, but the potential benefits were so significant that every manufacturing company needed to attempt the transformation if it hoped to remain competitive. The companies that succeeded were able to slash their inventories by half or more. The reliability of customer deliveries improved from percentages less than 70 percent to 95 percent or greater at most companies. The cost of materials management overhead was cut in half or more. And the cost of goods sold declined.

It took most of the 1970s for most companies to fully implement MRP, but it became the minimum level of performance to remain competitive. Managing with information replaced management based on physical inventory. Companies that did not make this transformation were no longer manufacturing.

It's worth noting that it took every company some time to make the MRP transformation to an entirely new generation of manufacturing management. Each of the individual systems within MRP needed to be implemented. The quality of the data needed to be improved. And finally, a broad range of new manufacturing processes and practices had to be implemented to achieve the benefits. Implementing this new generation was a journey, not an event, and this lesson applies directly to the next generation of product development we'll be considering in this book.

The Integrated Accounting Example

Accounting systems and processes provide a similar example. Prior to the 1970s, accounting systems did not integrate information, and the accounting process had to compensate for this. Accounts-receivable and accounts-payable information, purchasing information, and general accounting information were all separate and distinct. As a result, accounting transactions were very slow, many errors were made, and a great deal of manual work was required.

For example, a vendor invoice had to be matched manually to a purchase order and a receiving report, and the information from the three documents had to be manually reconciled. When this vendor payment was recorded, it had to be summarized by general ledger account number along with other payments, and then posted to the general ledger. A similarly inefficient process was necessary for accounts-receivable invoices, fixed asset purchases, payroll processing, and so on. Most companies had a department for each of these functions, such as an accounts-payable department, an accounts-receivable department, and a payroll department.

In the late 1960s, companies began to automate each of these individual accounting processes using new application software from providers like McCormack and Dodge. Some companies developed their own application software for these accounting processes, but most eventually decided that these were too expensive to develop internally. By the late 1970s, enough individual applications were available that integration of all accounting applications became the next objective.

By the late 1980s, integrated accounting systems and processes had become the standard. Now, when a vendor invoice is received, it is automatically matched to the purchase order and receipt, routed for approval, and posted to the appropriate general ledger account. Errors, delays, and inefficiencies have been removed from the process. Instead of separate accounting departments, most companies simply have one, with a greatly reduced accounting staff. Large companies today could not keep up with their accounting requirements without integrated accounting, because the number of transactions would require armies of accountants.

A similar transformation will take place in the next generation of product development as took place in MRP and integrated accounting. Isolated islands of information will be replaced by integrated information systems and integrated processes, which will enable a whole new approach to R&D management.

The Operation Iraqi Freedom Example

Increased information enables almost anything to be significantly improved, including military strategy, and Operation Iraqi Freedom in 2003 provides a dramatic example. The coalition forces had an overwhelming technical advantage in this conflict, but in the broadest sense the primary advantage was informational. Commanders on the battlefront received real-time images and targeting data from satellites, drone aircraft, and navigational systems. This helped to shrink the time lapse between identifying and hitting targets from hours to minutes. According to General Peter Pace, Vice Chairman of the U.S. Department of Defense's Joint Chiefs of Staff, speed was the key to success. "The ability to do things inside the enemy's decision cycle is an important strategic advantage. Force equals size times speed. Getting three divisions to the right spot in 30 days can be superior to getting five divisions there in 90 days."

That ability enabled the coalition forces to fight the war in Iraq with less than half of the forces used to liberate Kuwait in 1991, which was a considerably simpler operation. Operation Iraqi Freedom was successful, with fewer than 160 coalition soldiers killed and more than a third fewer civilian casualties than in the Gulf War.

The ability to destroy military targets has consistently improved with increased information and guidance. In WWII, as many as 3,000 bombs were needed to destroy a single target. In Operation Desert Storm, 10 planes were needed per target. Today, a single plane can cover 10 targets. "This was the first war that used fully integrated information. The synergy was enormous, and trust came from training as a joint force," said General Pace. "Connectivity helped us to understand the battlefield. The picture of the battlefield in the Pentagon was the same as seen by General Franks, and those on the ground in the battle. While it did make those on the ground worry about interference by generals not in the battle, the advantages were greater. Commanders understood precisely what they were talking about. Knowledge of the battlefield was complete."

Here again, information was used to make the process more efficient, and information replaced the need for physical assets. In fact, information technology has now enabled a new generation in the strategy, tactics, and processes of war.

CHARACTERISTICS OF THE R&D PRODUCTIVITY GENERATION

As discussed earlier, the next generation of product development management enables new competitive strategies based on greater productivity to create and bring to market more products, similar to the way the previous generation enabled time-based competitive strategies. In this next generation, companies can achieve extraordinary improvements in the productivity of their development teams, enabling a higher financial return on every project. They can dramatically increase the use of scarce resources and reduce resource costs simultaneously. They are able to develop more new products than before at the same or even lower costs. Product development can be closely integrated with product strategy, and companies can optimize revenue and profit from their new product pipeline. Finally, they can achieve financial control and financial integration of R&D for the first time.

Before we discuss the specific benefits of the R&D Productivity Generation, let's look at an overview of it. Each of the three sections of Figure 2–2 corresponds to sections of this book, and we'll spend time examining each of them throughout the book, but as a preview, it's useful to give an overview of each of them. The primary emphasis of the top section is strategic, while the primary emphasis of the two lower sections is managing the execution of strategy.

F I G U R E 2–2

Overview of R&D Productivity Generation Processes

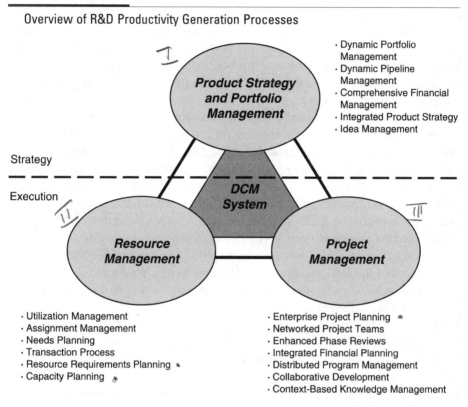

- Dynamic Portfolio Management
- Dynamic Pipeline Management
- Comprehensive Financial Management
- Integrated Product Strategy
- Idea Management

Strategy

Execution

- Utilization Management
- Assignment Management
- Needs Planning
- Transaction Process
- Resource Requirements Planning
- Capacity Planning

- Enterprise Project Planning
- Networked Project Teams
- Enhanced Phase Reviews
- Integrated Financial Planning
- Distributed Program Management
- Collaborative Development
- Context-Based Knowledge Management

Product Strategy and Portfolio Management

During the TTM Generation, most companies implemented some periodic management of the product portfolio after they achieved the advantages of improved project management. They collected summary information from all projects and began to analyze it strategically. Some also tried to do periodic pipeline management by comparing project priorities in order to aggregate resource availabilities and constraints. In general, these efforts at portfolio and pipeline management helped companies to somewhat better align their product development projects with their strategic priorities.

However, most companies quickly ran into limitations because the portfolio and pipeline analysis wasn't based on real-time actual project data. The data had to be collected, usually in summary form, and then consolidated manually. In the new generation, DCM systems will provide comprehensive portfolio and pipeline analysis directly based on actual project information. We'll look at how this will lead to additional advances in pipeline and portfolio management and what I call "on-demand" portfolio analysis. Section Four of the book addresses the strategic level of product development, including dynamic portfolio and pipeline management, comprehensive financial management, and integrated product strategy.

In addition to improved portfolio and pipeline management, two new major capabilities are being introduced in the next generation of product development. Comprehensive financial management of R&D consolidates critical financial information across all projects. Consolidated revenue and profit projections are created directly from the current financial projections of each project, as is other financial information such as consolidated capital requirements, the breakout of R&D investments, etc. Another important capability for financial management is the reconciliation of functional budgets with project budgets. Until now, CFOs were not even able to know whether 50 percent or 120 percent of the R&D budget was allocated to new products.

Product strategy is also ripe for new information systems, and this will encourage a more formal product strategy process. Capabilities in this vein include product line planning, product platform management, product lifecycle management, portfolio simulation, and management of new product ideas. Product strategy information can also be integrated with product development information, providing a powerful strategic engine. For example, future products included in a product line plan will provide longer-term visibility to pipeline and resource management.

Resource Management

Resource management is the management of a company's investment in development resources, primarily its people, in order to achieve the optimum product development output consistent with its strategy. As indicated in Figure 2–2, resource management primarily focuses on execution. The activities related to strategic resource allocation are performed at a higher level in pipeline management.

As you will see in Section Two, resource management starts with the management of project assignments, which enables short-term utilization management and some early productivity benefits. Resource management provides new capabilities for project managers and resource managers to improve the utilization of resources. Project managers will be able to plan projects based on the availability of critical resources. Resource managers will be able to avoid overloading their people while optimizing their utilization across all projects.

Utilization is one of the key concepts in the next generation of product development. Increased utilization improves R&D productivity and reduces the hourly cost of every developer. While utilization has been one of the most important concepts in other industries and functions, its importance in product development is only now being realized, so we will spend some time examining it. Utilization reporting, and the ability to manage R&D utilization, constitutes a critical underlying capability in development resource management. In the next generation of product development, companies will set utilization objectives and then manage to them in order to increase productivity.

> Utilization is one of the key concepts in the new generation of product development. Increased utilization improves R&D productivity and reduces the hourly cost of every developer.

There is a continued opportunity beyond the short-term management of resource assignments. Most companies will move beyond that stage to medium-term resource capacity planning and management. This will extend the visibility of resource needs beyond assignments, enable even greater utilization improvements, and help companies maneuver through the problems of resource constraints. We'll also look at the resource management transaction process, where project managers identify needs and request assignments, and where resource managers fill those needs and initiate adjustments where necessary. This is a cumbersome and inefficient administrative process in most companies today, but can now be automated and made far more efficient.

We will then shift our resource management focus from resource utilization across all projects to the more efficient use of resources within individual projects. How can project managers plan and schedule resources more efficiently in order to reduce the cost of their projects? Resource management data provide the basis for better planning and scheduling, but some new tools and techniques are needed as well. This is referred to as resource requirements planning and management.

Resource management was not comprehensively addressed in the Time-to-Market Generation. Project managers tended to assume that resources were infinitely available when they formulated their project plans, because they had no view of actual resource availability. Many companies did not even have a formal process for keeping track of who was actually working on what project. This means that there is a lot to be done to implement resource management in this new generation, but it also means there is an extraordinary opportunity.

Project Management

Project management has been improved in the last two generations of product development management, but it will be greatly improved in the new generation, and this improvement will significantly increase project productivity. Better project management starts with an important change that underlies all of the other improvements: its transformation from a technique to a management process. As a technique, the success of project management depends on the individual project manager's particular technical skills. As a process, project management becomes a systematic and clearly defined way in which a group of people work together to accomplish a common purpose. It coordinates the work of multiple people. This new project management process, which will be referred to as enterprise project planning and control, enables many new capabilities that were not previously possible.

The cross-functional core team was one of the most important improvements in the TTM Generation, but the core team structure has some limitations. I also strongly believe that the effectiveness of core teams was undermined in some companies by the out-of-control use of email communications. The next generation of product development enables an entirely new project team model that overcomes these limitations and inefficiencies. With the networked project team organizational model, team members will seamlessly coordinate, collaborate, and communicate for greater efficiency.

Phase-review management is also enhanced with an enterprise project planning and control process in place, and this process provides some significant benefits for executive management teams. They will have visibility of resource impacts when they make project decisions, and can quickly explore alternatives. Once they establish project tolerances for the project team, they can expect that these tolerances will be automatically monitored by the DCM system. Executive management teams will also have their own information system to make their work more efficient, give them more visibility into projects, and help make them more cohesive as a virtual executive team.

Financial planning and project budgeting for new products is increasingly important. Fortunately, the integration of financial information, along with standard financial models and new financial techniques, is making this planning and budgeting much more effective. Financial information can now be derived from operational information. Standard financial models can be used by all teams to increase their productivity and consistency. And, most importantly, with an integrated system, financial data are automatically consolidated across all projects. This enables improved financial management at the operational and strategic levels, to the extent that, for the first time in most companies, the chief financial officer is gaining financial control over R&D.

Enterprise project planning and networked project teams provide a foundation for additional advances in project management. Instead of project management being restricted to individual projects, a group of related projects can now be managed in a coordinated way. I'll refer to this capability as distributed program management, and we'll look at how enterprise project planning and control, networked project teams, and integrated financial management make this an exciting management alternative.

> Enterprise project planning and networked project teams provide a foundation for additional advances in project management.

This foundation also enables more effective management of collaborative development. We'll look at how collaborative development can be managed across a range of collaboration categories, and we'll also look at the collaboration system services required. Collaboration will eventually extend to the outsourcing of some R&D functions, so we'll look at managing this as well.

R&D is basically a knowledge activity, and there is a range of knowledge and experience that can be better institutionalized and leveraged to increase productivity. In the final chapter of the section on project management, we'll look at R&D knowledge management and introduce the concept of context-based knowledge management, which is also enabled by enterprise project planning and control. Context-based knowledge management is the delivery of the appropriate information and experience to the developer who needs it, when he or she needs it.

BENEFIT MODEL OF THE R&D PRODUCTIVITY GENERATION

Understanding the financial benefits of any major new initiative is essential, particularly after some questionable systems investments of the 1990s. The next generation of product development is not an exception to the need for clear benefits. These benefits do not necessarily need to be simple to understand, but they must be significant and real.

Early on in the TTM Generation, the benefits of improved time to market were estimated by projecting the revenue increase that could be achieved by bringing products to market sooner. To do this, companies first estimated their potential to improve time to market, for example, by 30 to 50 percent. It turned out that precision in this estimate didn't matter much, since the benefits were significant at almost any improvement. Then they projected what the new revenue curve would be if products were brought to market earlier, and estimated the revenue and profit benefit based on the new revenue curve.

A similar benefit-estimating technique can be applied to understand the potential improvements in the next generation of product development management. As you will see, the potential benefits are extraordinary, and I've constructed the benefit model in Figure 2–3 to help estimate them. The model assumes a current baseline of financial performance and then computes the financial impact of improvements to R&D productivity.

For the baseline, we will assume that a company invests $100 million R&D per year, and that this R&D investment will create new products that collectively result in $150 million of profit. After deducting the R&D investment required to develop the products, the net profit is $50 million, for a simple return on investment of 50 percent. Using the benefit model, you can later insert your own data, make your own assumptions, and compute your own expected benefits, but let's walk through the example. In it, there are seven per-

F I G U R E 2–3

Benefit Model for the R&D Productivity Generation

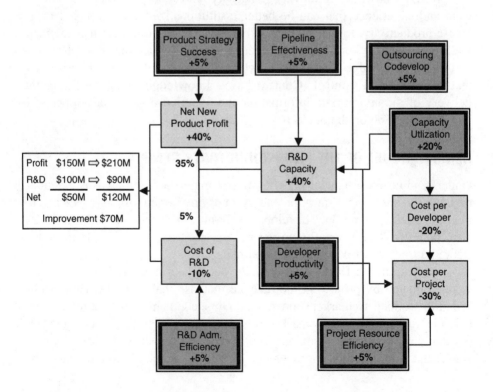

formance factors that drive increased productivity in this new generation, and each of these is shown in a highlighted box.

Improved R&D Capacity Utilization

Let's start with one of the biggest opportunities for improvement, and certainly the easiest to measure. Because R&D capacity is generally fixed, not variable, improved utilization of that capacity can create significant productivity increases. The importance of capacity utilization is well understood by executive managers of capital-intensive factories, airlines, and hotels, as well as by executives in service businesses, such as consulting firms, yet most companies whose success hinges on new product introductions have yet to take advantage of the benefits from improved R&D capacity utilization.

The objective of R&D capacity management is to get more new product output from a relatively fixed level of input, and the benefits of succeeding can be very substantial. For example, if a company has assigned 60 percent of its R&D resources to approved revenue-generating product development projects, and can increase this assignment to 80 percent, then it will achieve a 33 percent [20% / 60%] productivity improvement. As we will see in Section Two, the opportunities for improving utilization are much more significant than most companies realize. Moreover, companies can achieve a major portion of this productivity improvement relatively quickly, and then solidify and extend the benefits in subsequent stages of improvement.

The benefit model assumes a relatively conservative 20 percent improvement in R&D capacity utilization. This effectively increases development capacity by 20 percent, with an equivalent increase in the number of new products that can be developed. Companies can decide to take advantage of this improvement by reducing R&D investment while maintaining the same output, by increasing the number of products developed while maintaining the same investment, or by combining the two. As we will see in Chapter 3, higher utilization also reduces the hourly or daily cost per *applied* developer. In our model, this 20 percent improvement also reduces the cost per developer by 20 percent, which in turn is a major factor contributing to the expected decrease in project cost.

Increased Project Resource Efficiency

In the new product development generation, resources will be used more efficiently. Projects will be completed with fewer developers assigned to them, thanks to advances in project resource planning. Advanced top-down project planning, the ability to integrate that planning with resource availability, and the capability to integrate plans among projects will arm project team leaders with advanced capabilities to more efficiently plan project resource needs. Team leaders will find that they can accomplish all project objectives with fewer resources, thereby using scarce R&D resources more efficiently. I realize that there are skeptics who think that staffing projects more efficiently will increase time to market. As you will see, however, there is an opportunity to fulfill—if not exceed—the time-to-market objectives of the previous generation while at the same time using resources more efficiently.

> In the new generation, resources will be used more efficiently. Projects will be completed with fewer developers assigned to them, thanks to advances in project resource planning.

The benefit model assumes a 5 percent increase in project resource efficiency, but I believe that this is conservative. Most projects can achieve a 10 percent or greater improvement, but, unlike the utilization improvement that benefits all of R&D, project resource efficiency improvements are achieved project by project. If only half of the projects put these new practices to use, then a 10 percent benefit on those projects translates to an overall benefit of 5 percent.

Developer Productivity

Almost everyone agrees that there is still too much time wasted by developers on unnecessary or nonproductive activities, and, as we will see with the new project management practices described in Section Three, project team members will be much more productive in this new generation. They will waste less time on non-value-added project activities, leverage institutional experience, and work together more effectively. The benefits of increased developer productivity will vary by company, but I wouldn't be surprised to see most companies achieve at least a 10 percent improvement in all projects applying these practices. A 5 percent improvement in developer productivity is assumed in the model. Inefficient companies can reap even greater improvement-based gains.

Increased developer productivity means that projects can be completed in fewer, more productive hours, reducing the cost per project. In the benefit model, we see the significant cumulative impact from the reduced cost per developer, increased developer productivity, and improved project resource efficiency. Lower project cost can have a major strategic impact on product development by increasing the return on investment of all projects and increasing effective development capacity even further. We'll summarize this a little later.

Gains from Partnering/Outsourcing

Most experts believe that increased collaborative development, partnering, and outsourcing of selected R&D activities are inevitable. Just like manufacturing in the 1970s, R&D is highly vertically integrated, providing most of the resources needed through internal staffing. As was the case with manufacturing organizations, R&D organizations will discover that there are significant cost-reduction opportunities from reducing dependence on "doing everything yourself." The need for some R&D capabilities, such as product testing, fluctuates throughout the year, and productivity is reduced if a company permanently staffs these capabilities for peak demand to avoid slowing time to market. Frequently, the demand for specific skills varies widely based on requirements at any given time in the portfolio of projects. In still other areas, skill requirements are not high enough to justify maintaining world-class competencies, and outsiders can do the work better and more cheaply.

The productivity benefits from increased partnering and outsourcing will vary widely from company to company. In part, the benefits will depend on the level of efficiency that companies achieve in capacity management, since outsourcing is more cost-justified where R&D functions are less efficient, as was the case in manufacturing. So improvements in utilization may reduce the magnitude of cost-reduction opportunities through outsourcing and partnering.

There are three categories of benefits from outsourcing and codevelopment. The first benefit of collaborative development is a lower project cost. This reduction is attributable to portions of the project being performed by development partners for a lower cost, because of either their specialized expertise or their cost structure. The model assumes no benefit in the reduction in project costs from outsourcing and collaborative development. Outsourcing can also be used to reduce the cost per developer by doing some development in lower-cost countries or with lower-cost contractors. For the sake of simplicity, however, the model assumes no benefit here also. The third category of improvements from collaborative development is an increase in the number of projects completed, due to the flexibility advantages of collaborative development. The model assumes a 5 percent benefit, but improvements in effective capacity from collaborative development could be more significant in many situations.

There is yet another benefit from collaborative development that is less direct. By using external development partners or suppliers who share in the new product investment, a company can do more development with less investment, and later share some of the new product profit. While this opportunity can be strategically significant to some high-growth companies, it's not included in the benefit model because it's more of a timing advantage.

Increased R&D Administration Efficiency

Those intimately involved with product development know that too much time is spent on non-value-added management and administrative tasks, such as assigning people to projects, scheduling management reviews, collecting data for portfolio reviews, coordinating across projects, and so on. These activities are distributed across an organization and can be significant when added together.

In the new product development generation, administration and management will be much more effective. More R&D management and administration will be accomplished at a much lower cost. In the benefit model, this improvement is reflected as a 5 percent increase, but it could be much higher for some companies.

Pipeline Effectiveness

Product development pipeline management will be greatly improved as well. Better resource management, improved strategic management of R&D, and real-time integration with project information and status will enable much better response to changing conditions. In fact, many of the practices we will review will benefit pipeline management, either directly or indirectly. The benefit model assumes a 5 percent improvement in pipeline effectiveness, translating to a 5 percent increase in project completions.

Some companies may also obtain a benefit from gaining the ability to cancel problematic projects earlier in their development cycles, thus avoiding wasted development. This savings is not included in the benefit model, since it is assumed that these benefits have already been fully realized in the previous generation of product development. This may not be the case with some companies, however, in which case this benefit should be added to the model.

Improved Product Strategy and Product Success

This category includes a variety of improvements, not the least of which are better products that are more successful in taking advantage of market opportunities. Some of the new or improved practices, such as more effective idea management, increased collaboration with customers, and greater leverage of institutional experience, will lead to improved products and increased revenue. While this benefit is difficult to quantify and predict, in the end it could be the most significant benefit for some companies.

The benefits of improved product strategy, the introduction of financial management of R&D, and a better product development mix are also difficult to quantify, but could be an important improvement for some companies. In almost all companies today, R&D is not effectively aligned with business planning and financial management. Because all the R&D project information is scattered on desktop computers and maintained in inconsistent formats, companies are unable to consolidate critical planning information, such as the planned revenue from new products currently being developed. As a result, they don't realize that the expected revenue from projects in development could be much higher—or, even worse, much lower—than what they are planning. Correcting this problem could be vital to a company's very survival, so the benefit could be infinite.

> In almost all companies today, R&D is not effectively aligned with business planning and financial management.

The benefit model arbitrarily assumes a 5 percent benefit to new product profit through better products, better product strategy, and improved product mix in the next generation of development. For most companies this is an intangible benefit, but one that should be considered in the benefit model.

Resulting Benefits

Let's summarize the benefits of the next generation of product development, as depicted in the model. The cost per applied developer is reduced by 20 percent due to an increase in capacity utilization. A reduced cost per developer provides significant cost advantages, and enables a company to better compensate developers while at the same time maintaining a lower cost than competitors.

The cost per project is reduced by 30 percent! This is due to the 20 percent reduction in cost per developer, the 5 percent cost decrease due to developer productivity gains, and the 5 percent decrease due to more efficient resource use. We assumed no reduction of project cost from outsourcing.

It's important to understand the impact of project productivity on the profitability of new product investments. Let's take the example of only a 15 percent productivity improvement on a $4 million project with a projected profit of $5 million. After deducting the project cost, the net profit for this project is $1 million [$5M – $4M], for a return on investment of 25 percent [$1M / $4M]. However, if the project cost is reduced by 15 percent, to $3.4 million [$4M – $4M × 15%], then the net profit would be increased to $1.6 million [$5M – $3.4M], and the return on investment would rise to 47 percent [$1.6M / $3.4M]. For some companies, this increase in project productivity will mean that project opportunities that previously were not good investments now become excellent ones.

R&D capacity is increased by 40 percent due to the 20 percent improvement in capacity utilization, the 5 percent increase in pipeline effectiveness, the 5 percent capacity improvement from outsourcing, the 5 percent improvement in developer productivity, and the 5 percent improvement in project resource efficiency. A company can choose to translate this benefit into more new products from the same investment, a lower R&D investment with the same output, or some combination. The model assumes that 35 percent of the 40 percent improvement is applied to increase the number of products developed and 5 percent is applied to reduced R&D investment. Most companies may apply the benefit in this way, since it's difficult to quickly reduce the number of developers. The 35 percent increase in the number of new products directly increases the profit from new products, since more projects can be completed annually, and when this profit increase is combined with the 5 percent improvement from better product strategy and product mix, the result is a 40 percent improvement in new product profit. The 5 percent reduction in R&D investment combined with the 5 percent improvement in R&D administrative efficiency results in a 10 percent reduction in the cost of R&D.

In total, the benefit model depicts extraordinary results. The cost of R&D is reduced by 10 percent, from $100 million to $90 million. This reduction could have been much greater, but the model shows that most of the reduction is being re-invested in more product development. New product profit increases by 40 percent, from $150 million to $210 million. This increase reflects the profit potential of additional products, and therefore assumes that a company has additional valuable new product opportunities that it can pursue. If not, it would do better to further reduce its R&D spending instead of increasing its effective R&D capacity. The net profit is now projected to be $120 million [$210M – $90M], instead of $50 million—for an overall improvement of $70 million.

You can imagine the strategic impact that these benefits will have on competitive advantage and return on investment. Companies that can develop products at a 30 percent cost advantage compared to their competitors can flood the market with more new products because they cost less to develop. By getting more from its R&D investment, a company can increase its profit margin with lower R&D expense while at the same time increasing growth through greater R&D capacity. In short, it can do more R&D for less.

DEVELOPMENT CHAIN MANAGEMENT SYSTEMS

The next generation of product development described in this book is based on integrated software systems that automate information flow, visibility, and decision making to enable an entirely new level of R&D performance. I'll refer to these systems generically as development chain management systems, or DCM systems, even though there is not yet a broadly accepted term used to describe them. There are a number of software applications that generally fit into the general scope of DCM systems. Some of these applications automate portions of the broader DCM framework we will be applying in this book, while others provide more complete solutions.

It is important to note a few distinctions with regard to DCM systems. These systems are different from software tools, such as project planning tools, used by individuals. DCM systems are integrated across the enterprise, so they are network-based applications, not desktop applications. DCM systems are different from operational PLM (product lifecycle management) systems because DCM systems automate the management or project layer of product development, not the operational or product layer.

Figure 2–4 shows the systems that comprise an integrated DCM system, along with the lower PLM operational layer. As you read through this book, the purpose of these systems will become clearer, and you may want to refer back to this figure periodically. The DCM systems it depicts are grouped in a similar way to the major areas of product development that we'll be discussing in detail in Sections Two to Four. There are three groups of systems: resource management systems, project management systems, and product strategy systems. Each of these groups, in turn, includes a number of applications or subsystems that work closely with one other. The application names and integration features vary with different DCM systems.

Resource management systems manage and optimize project resources. An application for managing the assignment of resources to projects is at the center, because this is the most basic resource management capability. Surrounding this resource assignment application may be applications for medium-term resource management that support planned needs and capacity management, utilization reporting, integrated tools that support resource requirements planning (RRP) for project managers, and an integrated transaction-management process for project and resource managers.

F I G U R E 2–4

DCM Systems Integration with PLM Operational Systems

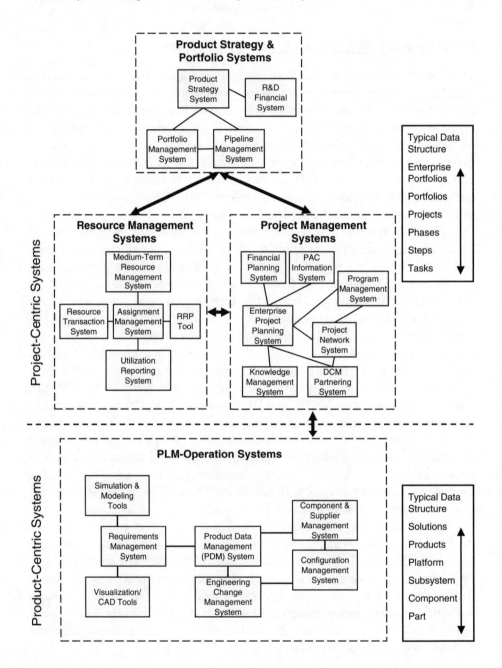

The project management system broadly includes a range of applications that support the advanced project management practices of this new generation. Enterprise project planning systems and project network systems are the most critical applications, since they provide the foundation for DCM. Other project management applications that leverage this foundation include an integrated financial system, an information system for executive management, a program management system for companies implementing program management in addition to project management, an integrated partnering system for collaborative development management, and an integrated knowledge management system.

The system defined as product strategy and portfolio management includes advanced applications for online pipeline management and portfolio management that are integrated with project data in the DCM system. It can also include an R&D financial system that provides integrated financial management by consolidating and analyzing financial information across all projects. Finally, some companies will eventually advance to applications for product strategy that will be integrated within the DCM system.

These systems are not hypothetical. Several software companies now offer systems that include most or all of these applications. What is distinctive in our discussion of the next generation of product development practices is how they enable the kind of integration that end-to-end supply chain management enables. That's why I call them development chain management systems.

COMMON CONCEPTS

Like any major new body of knowledge, the next generation of product development introduces many new concepts and frameworks, as well as a lot of new terminology. Many of these terms are defined in the glossary, and you may find it useful to refer to it if you are confused about how a particular term is used in this book. A few frameworks are used consistently in each of the three major sections, and it may be helpful to introduce two of these before going into them in more detail.

Management Process Levels

Management processes consist of multiple levels, with each lower level managing progressively more detail. The emphasis tends to be on strategic management at the higher levels, and on tactical management at the lower levels. Higher levels usually are managed with more aggregation of information, and the lower levels in a more "granular" fashion. These distinctions are important in understanding different aspects of a complex management process and seeing how they map to organizational levels.

These management process levels will be used to organize various practices in each of the three major sections of the book, but it's helpful to introduce this general concept of higher-level strategic and lower-level tactical management up front to provide the context. For example, in project management, the top level is the project phase level, which is used for phase-based decision making and tracking projects strategically. The middle level is the step level, where project managers focus their attention on planning and control of their projects. The lowest level is the task level, which is the focus of the work done by developers.

As depicted in Figure 2–5, there is a three-level framework for each of the major parts of the next generation, and these levels are used to combine the major processes. For example, project resource assignment management is directly integrated at the phase level with projects. This integration of processes will become clearer as we explain it in more detail, but you may want to periodically refer back to the relationship of the levels in this framework.

F I G U R E   2–5

Levels of Product Development Management

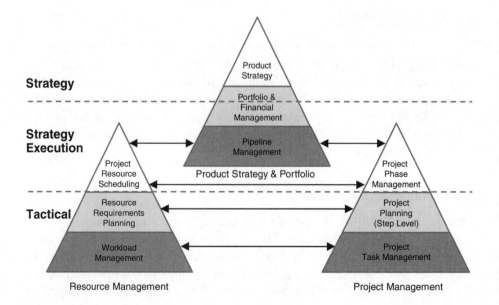

Stages of Maturity

Stages-of-maturity frameworks have been used with great success to break a complex management process into groups or stages of management practices. Seeing how they evolve makes them easier to understand because they can be explained in sequence. Stages-of-maturity frameworks also provide a useful roadmap for implementing new management practices and, most importantly, define the benefits a company can expect along the way. Finally, a stages-of-maturity framework enables companies to benchmark their current level of performance relative to others, and explicitly identify potential gains.

> Stages-of-maturity frameworks provide a useful roadmap for implementing new management practices and define the benefits a company can expect along the way.

We've defined a stages framework for each of the three major areas of the next generation of product development, each showing where most companies stand at the tail end of the Time-to-Market Generation. These three stages-of-maturity frameworks can also be combined to guide an overall implementation of this new generation of product development management. For example, a company can progress to one stage of project management, and then focus on advancing to the next stage of resource management.

COMMERCIAL ROBOTICS INCORPORATED

The R&D Productivity Generation that is just emerging, and the practices described in this book, are still so new that it is very difficult to find sufficiently meaningful real-life examples for purposes of illustration. Even though many of these new-generation practices are now in use, few companies, if any, have fully deployed them. Additionally, companies that are leading these practices are reluctant to talk publicly about them, for competitive reasons.

Yet examples are a powerful way to illustrate these new practices in action, so I've created a hypothetical company, which I'll use to broadly demonstrate the practices and expected benefits of the next generation. While hypothetical, this company represents a composite of customer experiences with the current DCM systems. I named this company Commercial Robotics Inc. (CRI), and it designs, manufactures, and markets robots for use in retail, wholesale, and industrial applications. As the leader in this rapidly growing industry, CRI has been growing by more than 25 percent per year. It currently

has approximately $1 billion in revenue and invests approximately $100 million annually in research and development.

Typical of executives at many rapidly growing companies, Brian Kennedy, the chief executive officer of CRI, is recognized for his vision and ag-
gressiveness. As you will see, he believes that CRI has unlimited opportunities
for new products, and that improving R&D productivity is an important competitive advantage.

Chris Taylor is the VP of R&D, whom Kennedy brought in to take the
company's product development to the next level. Taylor wants CRI to be the
best in the world at developing new products, and wants to have the best product development process. We will see how CRI achieves this ambition as we
progress through the book.

Shaun Smith is the new CFO at CRI. He was brought in by the board of
directors to ensure financial control of CRI as it grows rapidly. As you will
see, he has a particular interest in improving the financial management of
R&D, similar to other CFOs today.

We will also look closely at one of CRI's most strategic projects, the
Fast-Food Robot. This is a robot that will interact with customers at ma-
jor-chain fast-food restaurants, taking customers' orders and getting their food.
We will see how the Fast-Food Robot project team applies enterprise project
planning and control, uses the networked project team organization, performs
integrated financial planning, and collaborates with development partners.
Anne Miller, the Fast-Food Robot project manager, is one of CRI's strongest
project managers, and eagerly volunteered to be the first to fully apply the
practices of this new generation.

The Fast-Food Robot example, like the company name and its employee's
names, is entirely fictitious, as are the business activities referenced between
CRI and real and imagined customers and suppliers. I'm not aware of any such
products currently being developed.

While I talk about the characteristics of the new or next generation of
product development as if it is already at hand, I realize that for many readers
it is not. But the vision that I outline in this book is based in reality, not con-
jecture. The hypothetical case that I use to illustrate the practices of the new
development management generation is based in part on real experiences in
using advanced systems available today.

Resource Management

3
CHAPTER

Improving R&D Productivity through Increased Capacity Utilization

The management paradigm of the previous generation (the Time-to-Market Generation) was the individual project, and everyone recognized that the development cost of a project was not as important as getting it to market fast. Even if additional resources were necessary, the benefits of earlier revenue, a longer product lifecycle, and competitive advantage almost always outweighed any additional costs. So the new practices of that generation primarily optimized time to market. Each project was planned independently, and project managers generally assumed that infinite resources were available when they planned their projects. By starting every project step as soon as possible, risks of delay were minimized. Senior management made crisp Go/No-Go decisions at just the right time and empowered the project teams to execute the next phase. Yet by focusing almost exclusively on optimizing the speed-driven returns on individual projects, companies compromised the efficiency with which they used their precious resources, and thus their overall R&D productivity.

The next generation of R&D management expands the focus to maximizing the overall productivity of development resources, thus looking beyond both projects and project speeds in isolation. At the same time, it builds upon the practices and benefits of the TTM Generation, enabling a better balance between the individual project and R&D productivity as a whole.

R&D productivity is an untapped opportunity for most companies, and it's a primary focus of the next generation of product development. In this section we'll look at how resource utilization affects R&D productivity, both at the project level and, even more importantly, across all projects. We'll also look more broadly at capacity management, which includes planning and managing the necessary resources in addition to utilizing them. R&D is predominantly a fixed-capacity operation, and increased capacity utilization directly improves productivity. In the next generation, advances in R&D capacity management are enabled by development chain management systems and new management processes and practices, creating a new paradigm for R&D management.

R&D is predominantly a fixed-capacity operation, and increased capacity utilization directly improves productivity.

Capacity utilization is hardly a new management principle. Businesses ranging from airlines to hotels to factories apply it routinely. Capacity utilization is also *the* most critical management principle in professional service businesses such as consulting, accounting, and legal services, where it is a key competitive factor. What we'll be exploring in this chapter is a perspective on R&D as a fixed-capacity operation, although it has not traditionally been managed that way. We'll explore this fixed-capacity perspective for the purpose of finding a new way to manage R&D for a greater yield of innovative new products from limited R&D resources.

It's easy to understand that R&D investment (capacity) is generally fixed, with very little variability in any given year, since the resources used to develop projects primarily consist of a company's internal R&D staff and other internal functions customarily involved in development. For example, if a company invests $100 million per year in R&D, most of that investment, let's say $90 million, is fixed, with the remaining $10 million variable. It doesn't really vary based on demand for new or improved products. Even if the company doesn't do any product development in a given year, it will still "invest" more than $90 million in R&D. On the other hand, if it does a lot of product development in a given year, its R&D investment spread across its portfolio of projects will still not significantly exceed $100 million. R&D investment, in other words, is largely insensitive to demand. The input—R&D investment—is relatively fixed; it's only the output—products developed— that varies.

What we'll explore here is how increasing the utilization of fixed R&D capacity affords an extraordinary opportunity for productivity increases. We'll see how, for example, if a company has 60 percent of its R&D resources assigned to *approved* product development projects (i.e., those projects that are part of its strategy), and can increase that figure to 80 percent, then it will achieve a 25 percent [20%/60%] productivity improvement. For some companies, the improvement may be even greater.

MEASURING THE OPPORTUNITY TO INCREASE CAPACITY UTILIZATION

Commercial Robotics Inc. (CRI), our hypothetical company, illustrates this opportunity. CRI spends approximately $100 million on R&D, with little of that budget contracted outside. It has 600 developers and related staff working on R&D, so its average fully loaded cost per developer per year is approximately $167,000 [$100M / 600]. Generally, the fully loaded cost per developer ranges from $150,000 to $200,000, factoring in benefits, facilities, overhead, travel, related expenses, etc.

Until recently, CRI—like most companies—didn't know its actual utilization of R&D. Chris Taylor, the company's vice-president of R&D, was very surprised to find out that it was only 60 percent. "I knew our R&D resources weren't aligned with our approved projects at any point in time. That simply made sense, since we have a wide range of skills, and it's not likely that you can match the mix of skills we have to the mix of skills we need at any point in time. My instincts told me that we could improve utilization. But with everybody around here working so hard, I thought it would be much higher than 60 percent. This tells me we have a lot of room for improvement, and it has nothing to do with getting people to put in more hours."

The 60 percent utilization of CRI's 600 developers is based on approved new product development projects. Specifically, these are revenue-generating projects. The remainder of the capacity is used for customer support activities, internal projects that are not involved in directly developing products, administration, developers who are available and awaiting assignment, developers who are less than fully assigned to projects, developers with additional time on their hands due to project delays, and so on. In addition, there's the time allocated for vacations, training, sick time, etc. Despite the feeling that everyone is busy, this is a typical profile for many companies. Some companies find they have better utilization and some find that they have much worse, but almost all share the perception that everyone connected with R&D is putting in long and hard hours.

CRI went through the calculations shown in Figure 3–1 to measure its opportunity for improvement. It first computed the simple cost per developer by taking its $100 million investment in R&D and dividing this by the number of developers, arriving at a figure of $167,000 per developer. It then computed

F I G U R E 3–1

Benefits of Increased Utilization at CRI

	Current	Improved	
R&D Investment (000)	$100,000	$100,000	
Number of Developers	600	600	
Cost/Developer (000)	$167	$167	
Utilization	60%	80%	+33%
Number of Applied Developers	360	480	+33%
Cost/Applied Developer (000)	$278	$208	−25%

the real cost per *applied* developer (the percentage of developers actually working on new product projects), which was much higher: $278,000 [$100M / (600 × 60%)]. This represented the average cost per developer applied to new product development, including the supporting costs and costs of available capacity. This works out to an effective cost of approximately $134 per hour [$278,000 / 2,080 hours] per developer.

CRI then measured the benefit it could obtain from improving its utilization from 60 to 80 percent. Instead of having 360 *applied* developers [600 × 60%], CRI would have 480 *applied* developers [600 × 80%]. This would effectively amount to a 33 percent increase in the company's R&D force, at no additional cost. Alternatively, CRI could maintain the 360 *applied* developers while reducing the developer headcount by 150, bringing its new total to 450 [450 × 80% = 360]. This would save the company $25 million [$167,000 × 150], or 25 percent of its R&D budget, while maintaining approximately the same productivity.

Brian Kennedy, the company's CEO, wants to reinvest the entire productivity increase back into R&D. "We have many more new product opportunities than we can pursue, and I want to reinvest this utilization improvement and get even more back in return. We consistently achieve a good return on our R&D investment, so we can reinvest this savings and make a lot more."

The reduced cost per *applied* developer was another way CRI looked at this improvement opportunity. With utilization increased to 80 percent, this cost would be reduced to $208,000 per year [$100M / 480] from the previously computed $278,000 per *applied* developer. This would also reduce the cost per developer per hour from $134 to $100. Chris Taylor thought this was very important. "The cost per hour will increasingly be a competitive factor. It's already beginning to drive outsourcing of R&D, especially the transfer of some development work to lower-cost countries. I want our development costs to be $100 per hour, while it costs our competitors $130 or more, and I want to do this at the same time that we're paying our people more than they're paying theirs. This means to me that we need to be more productive than our competitors."

The near-term opportunity to save or reinvest $25 million per year is significant for any company. In fact, most companies would put it on their "A List" of priorities. This benefit is also scalable. It applies proportionately for companies that invest more in R&D and for companies that invest less, although at very low levels of R&D investment there would be a smaller opportunity for utilization improvement.

In the longer term, higher R&D utilization creates a competitive advantage, just as it did in manufacturing. The lesson couldn't be plainer: In the 1980s, companies that increased their manufacturing capacity utilization achieved a competitive cost advantage. Those that didn't are no longer around.

The magnitude of this opportunity raises an obvious question: Why haven't most companies already increased their R&D capacity utilization? There are several reasons why this has not yet received a lot of attention. As stated at the beginning, the previous emphasis of R&D management was on individual projects rather than on resource management, and this emphasis has helped perpetuate the illusion that R&D investment is variable, based on the number and mix of projects approved. Product approval committees generally approve R&D investments project by project instead of by allocating fixed capacity among the approved projects.

Then there is the mistaken inference that if everyone in R&D is so busy, and there are never enough resources available, it goes without saying that capacity is being fully utilized. In fact, it's highly unlikely that all the diverse mix of skills of all the people in R&D coincidentally match the needs of the projects in a company's portfolio at any given time. It's also highly unlikely that utilization will somehow optimize itself without being explicitly managed. This is the same mistaken inference that manufacturing executives made before they began explicitly managing manufacturing capacity utilization.

Also, R&D executives have traditionally not been held accountable for R&D utilization. This is a relatively new performance metric, but one that is beginning to get more attention. Ask a VP of R&D, "What's your current level of R&D capacity utilization, and what should it be?" While these sound like obvious questions, few can answer them.

Finally, until now, R&D executives have simply not had the development chain management systems needed to manage resources and proactively manage R&D capacity utilization. This type of management cannot be done manually, or even by stand-alone tools. These crucial systems and the related processes are now available, and they enable this new paradigm for R&D management.

THE NEW PARADIGM FOR R&D: CAPACITY MANAGEMENT

Capacity management introduces a new management paradigm for R&D: It's the total output from R&D that is now the objective, not the time to market of the individual project. Capacity management now matters.

When R&D resource utilization is profiled factually, companies find some expected results. A portion of the resources, 10 percent for example, are significantly overassigned. Some people may be overassigned by as much as 300 percent of their available time, or even more. These gross overassignments are project delays just waiting to happen, but they rarely come as a surprise to anyone. But what is surprising to executives is the discovery that a significant percentage of their R&D resources is not fully assigned to approved R&D projects.

> Capacity management introduces a new management paradigm for R&D: It's the total output from R&D that is now the objective, not the time to market of the individual project.

What in the world are all these unassigned R&D people doing? The answer is that they're all busy, but they're not busy working on approved product development projects directly targeted to produce revenue. So what *are* they working on? Some are involved in activities other than product development. Customer support is usually high on the list. Others keep themselves busy by working on internal functional projects when not assigned to approved product development projects. While perhaps adding value in some way, these activities cannot be considered as utilization of product development resources. Companies that apply more of their R&D investment to developing new profit-generating products are much more successful than those that spend their R&D investment on internal projects. Finally, some developers use their unallocated time to help out on approved product development projects even though they haven't been assigned to them. While admirable, these self-assignments typically end up both increasing and masking the real cost of product development.

UNDERSTANDING R&D UTILIZATION

Achieving a high level of R&D utilization is not the same as being busy. Everyone in R&D is busy. High utilization means that an R&D organization is highly focused on its primary objective: developing products approved by the company's project-approval process. These are the primary investments necessary for a company to achieve its strategic objectives, and the primary purpose of R&D.

To understand utilization, let's look at the R&D utilization profile for CRI, as illustrated in Figure 3–2. In this example, the company discovered that only 50 to 60 percent of its R&D resources were assigned to approved product development projects. The percentage is expressed as a range because, in some cases, more than 100 percent of a person's time had been scheduled. This factor provided the range of 50 to 60 percent, depending on the likelihood of

F I G U R E 3–2

Typical Allocation of R&D Time

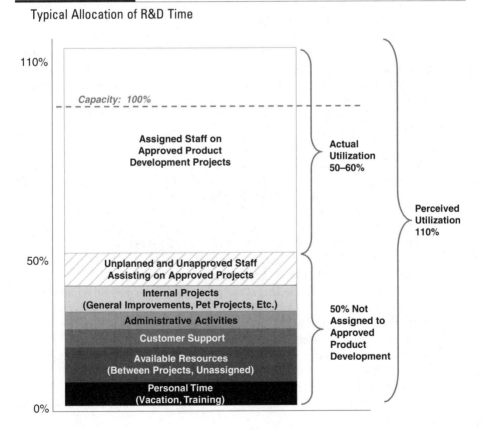

those overscheduled developers working the additional time, but it really should be measured as 50 percent. In this case, a number of CRI's developers were scheduled for more than 200 percent of their time and were not likely to fulfill their assignments. But fully 50 percent of the company's R&D capacity was not assigned to approved product development projects. Following is a profile of how CRI was using its R&D capacity.

Resources assigned to approved product development projects. Developing approved products is the primary objective of R&D, and this is the category of capacity utilization that should be optimized. The more a company's R&D resources are assigned to approved projects, the more approved projects will be completed and the more likely the company will be to achieve its strategy. This classification includes only resources assigned to projects at the time of their approval. It does not include resources applied to approved projects that were not included in the projects' resource authorizations.

Unplanned and unapproved resources assisting approved projects. Developers who have available time will frequently work on a project even though they are not assigned to it. This is laudable, and is often a good use of excess time, but this surplus time is a resource that should be available for assignment to another project. Note that the cost of this resource is not included in the project budget or the cost justification for developing the new product. It might be appropriate to determine if developers working on projects not assigned to them should be formally assigned to those projects. If so, then the project should be replanned to reflect the increase in its approved resources.

Unapproved projects. In some companies, product development projects start without approval and are hidden from the portfolio of active projects. These projects may or may not make sense strategically, but are "invisible" to a company's strategy, and any allocation of resources to them should be questioned.

Internal projects. These projects include all of the internal projects an R&D organization carries out to improve its capabilities, such as creating new development tools, advancing a new technology, improving the product development process, etc. Internal projects are not related to a specific new product, but can still be valuable. Frequently, these projects are carried out using unassigned resources, but sometimes are of sufficiently high priority that they compete for resources with approved product development projects.

Administrative activities. This is the amount of time spent on administrative responsibilities unrelated to specific development projects. These activities include resource administration, internal meetings, performance reviews, functional management, portfolio management, system administration, process management, and so on. The obvious objective is to reduce the time spent on these activities.

Customer support. Frequently, developers are called on to support customers, whether in a pre-sale situation, post-sale implementation, trouble-shooting, or problem resolution. These activities are essential, but are not really R&D, and their cost might be better reclassified out of R&D and into the cost of customer support. In some companies this allocation of R&D resources to customer support can be surprisingly high, and may call for a strategic reassessment of the R&D investment if too much of it is being channeled into customer support.

Available resources. These are idle resources not currently scheduled or only partially scheduled, and they represent an opportunity to launch another project or accelerate an approved project. Resources in this category can include:

- Developers who are between projects
- Developers who have skills that are currently not needed in the quantity that they are available
- Developers who are assigned to projects part-time, but have the other portion of their time available
- Developers who were assigned to a project recently canceled and have not yet been reassigned
- Developers who are not in demand by project managers for performance or other reasons
- Developers who are still working on approved projects beyond the time originally assigned, but have not been assigned for the additional time needed

Personal time. This includes vacation, sick time, holidays, leaves-of-absence, etc. Time for training, internal functions, and the like may be included in this category, or categorized separately.

MEASURING TRENDS AND GOAL SETTING

In designing a capacity management process, there are some alternatives to be considered. Capacity management starts by establishing a baseline for capacity utilization and then setting goals for improving it. Utilization trend reporting identifies progress over time and variations from goals. There are two different ways to measure utilization, and, as part of utilization management, a company needs to determine which one, or what combination of the two, is most effective. There are also multiple levels of utilization reporting, and a company needs to determine what levels of reporting make sense.

Trend Reporting and Goal Setting

Utilization trend reporting provides the basic information for capacity utilization management, as well as a way of measuring progress. CRI's utilization trend is illustrated in Figure 3–3. Over the first six months from, January to June, CRI established a utilization baseline. Utilization fluctuated around an average of 60 percent throughout that period. As previously mentioned, this

F I G U R E 3–3

Improvement Trend in Overall R&D Utilization for CRI

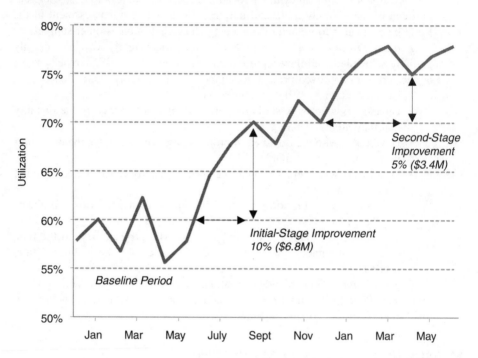

was somewhat surprising to the company's vice-president of R&D, who expected it to be higher.

During May and June, CRI implemented its first stage of resource management improvements, which delivered immediate benefits. These improvements involved adjusting staffing on projects, accelerating projects by using underassigned or unassigned developers, lowering the priority of internal projects in order to assign developers to revenue-generating projects, slightly shifting the mix of projects, delaying a couple of projects to better use resources, and approving the use of available developers on a couple of new, lower-priority, product development projects. These improvements were quickly implemented, and the results for the next three-month period (Initial Stage Improvement) were significant. Utilization increased by 10 percent (from 60 percent to 70 percent), which CRI equated to a $6.8 million per year productivity savings. As was its intention, CRI reinvested this savings to increase the number of new products developed.

CRI implemented the next set of resource management improvements (Second-Stage Improvements) over the next few months. These included capacity-based pipeline management, resource group assignment balancing, resource-based phase review decisions, resource-based project planning, and medium-term capacity and skill planning. Over the subsequent three months, these improvements contributed another 5 percent to utilization improvement, equal to a $3.4 million benefit, and the company expected continuing utilization improvement over the next six months.

This second stage of improvements is also critical because it makes the initial-stage improvements sustainable. It puts in place the necessary resource management process and systems. Without these, the benefits of the initial stage of improvements would diminish, until the company launched another temporary improvement project.

The CRI example illustrates how trend reporting measures progress in utilization improvement as well as quantitative measurement of the actions taken to improve utilization. It also illustrates how significant the benefits can be from better capacity management.

Basis for Computing Utilization: Assignments versus Actual-Time Utilization

R&D capacity utilization can be measured in two different ways. The first is by collecting actual time from all developers and summarizing how they spent their time during a particular period, such as a week or a month. This is typically done using timesheets or some automated variation requiring developers to record their time. While actual time information is valuable, there are several problems in collecting it. First, developers hate to record how they spend their time, making it very difficult to get compliance, let alone accurate data. Second, actual time collection is historical, meaning there's a time lag between getting all the time reports collected and analyzing them, making it too late to take timely corrective actions. Finally, the time-spent reports don't describe how developers were assigned to spend their time.

The alternative is to measure utilization through assignment information. Developers are considered to be utilized if they are assigned to an approved project. If a developer is working half time on approved projects, then he is 50 percent utilized, and if six of eight developers in a group are fully assigned to approved projects during a month, then that group is at 75 percent utilization. Basing utilization measurement on approved project assignments has the advantage of being the more accurate indicator of how resources are deployed. It is also the easier measurement to take, since it is a by-product of information required for resource management and doesn't require compliance from all developers—or the administrative enforcement of that compliance. Utilization measurement based on assignments creates a "snapshot" of how resources are assigned at a particular point in time; for example, at month-end.

This approach obviously has some inaccuracy built in, since developers can use their time differently from how they were assigned to use it. Some R&D executives argue that assignment information is still more reliable, since it measures how resources were "contracted," and if a project manager doesn't use the developer time assigned, the manager nonetheless "bought" that time. Others still want the accuracy that actual-time collection promises.

Eventually, the best practice is to use both measures together. Assignments measure the utilization of how resources are deployed; actual-time reports contrast actual deployments of R&D staff to the planned deployment. Since the biggest benefit comes from assignment-based utilization, and since it is a by-product of resource management, the assignment measure is usually implemented first, and actual-time reports are collected and analyzed subsequently, to measure actual utilization to assignments.

Levels of Reporting

Utilization reporting takes place at several levels beyond overall R&D capacity utilization. The most useful for utilization management occurs at the resource group level, since this is where primary responsibility for utilization is usually placed. Figure 3–4 illustrates the resource group utilization report used to measure the resource group's utilization against a pre-determined standard for CRI's Software Engineering Group. Year to date (YTD), the group's utilization is generally above the standard of 75 percent for the first six months. The standard is reduced to 60 percent for the summer months, and then returns to 75 percent.

In addition to measuring and reporting utilization by resource group, it may be useful to measure it by skill category and combinations of related resource groups, divisions, and business units.

In addition to measuring and reporting utilization by resource group, it may be useful to measure it by skill category and combinations of related resource groups, divisions, and business units. Utilization standards or goals can also be established for each level of resources managed, and then consolidated to determine overall goals for R&D. Standards can be based on experience, external benchmarks, or improvement objectives. As part

F I G U R E 3–4

Typical Utilization Chart for a Resource Group

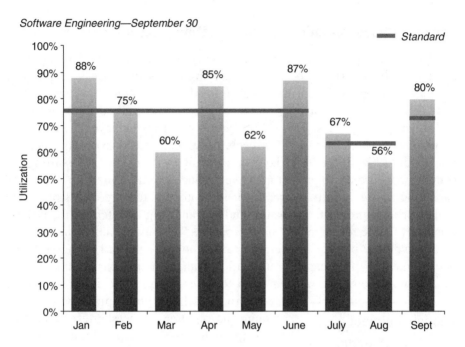

Software Engineering—September 30

of its capacity utilization management, a company should also establish utilization performance expectations for resource group managers. This introduces the critical issue of accountability: Who in the organization is accountable for utilization? Even recognizing that there are interdependencies between project managers and resource managers, it is still a good management practice to assign specific responsibility.

BEST PRACTICES FOR IMPROVING UTILIZATION

As I mentioned earlier, the Time-to-Market Generation applied innovative best practices to accelerate the completion of individual projects. While very effective at the time, these practices necessarily used simplifying assumptions, because comprehensive information across the length and breadth of the product development process was not available. These simplifying assumptions led to less than optimal results—results that can now be achieved with the new focus on capacity utilization management. This focus introduces a number of new best practices for increasing R&D utilization while maintaining the time-to-market benefits achieved in the previous generation.

Capacity-Based Portfolio and Pipeline Management

Portfolio and pipeline management in the previous generation focused almost entirely on trying to get as many high-priority projects as possible through the pipeline. The usual practice for doing this is to prioritize all projects and cancel lower-priority projects in order to allocate their resources to the higher-priority ones. This redistribution generally achieves the objective of getting the highest-priority projects developed, but doesn't consider the impact on utilization of the fixed investment in R&D. In fact, shifting resources from project to project may even reduce overall utilization. How? By canceling lower-priority projects to alleviate the resource constraints on higher-priority projects, companies end up canceling or deferring the utilization of available resources on these projects. These available resources don't shift to a higher-priority project; they are simply not utilized on new product development. With a capacity-utilization management focus, companies try to maximize the output of their fixed R&D capacity, while at the same time trying to get the highest-priority projects through the pipeline in the shortest time. This focus optimizes overall output through the best mix and timing of projects.

> Changing project schedules to better use a fixed-capacity perspective in conjunction with portfolio analysis will almost always increase development output.

The measure of overall output in this case is not simply a prioritized list; it is a set of alternative development portfolio scenarios for achieving the best expected revenue and profit. For example, a company could evaluate a number of alternative scenarios and compare expected revenue and profit by year for each scenario using the consolidated forecast from each project. Sometimes revenue and profit will be totaled across all years and used as the measure to be optimized. In other cases, revenue and profit could be risk-adjusted or adjusted for the present value over time.

Changing project schedules to better use this fixed-capacity perspective in conjunction with portfolio analysis will almost always increase development output. In some cases, companies may even initiate a new product development project in order to make use of available capacity, effectively getting that product developed for free.

Increased Awareness of Capacity Utilization

As you will see in Chapter 5, simply increasing the awareness of capacity utilization creates subtle changes that improve capacity management. Resource managers who are held accountable for the utilization of their resources will act more proactively to improve utilization. Project managers will be more careful in requesting resources, and will understand their responsibilities for capacity management.

This increased awareness was evidenced at CRI when a project manager decided to defer the start of an approved project. He had recently received approval for Phase 1, but wanted to delay it for three months because he was having difficulty getting sufficient time of his own to get it started. Two of the resource group managers asked him, "What are you going to do with the resources assigned to you?" "What do you mean?" he responded. They proceeded to tell him that once resources were assigned to a project, they were in effect being denied to other potential projects, which imposed an opportunity cost. They also pointed out that all resource assignments had an impact on the utilization of their own resource groups. Their message: "You should take responsibility for finding other projects for all the resources assigned to you." The project manager got the message, and decided to make the personal time to get Phase 1 of the project started as planned.

Resource-Based Phase-Review Decisions

In the previous generation of product development, many companies were able to significantly decrease project delays caused by initiating new development projects without assigning them the resources necessary. Although interpretation of the terms is not consistent, making sure the necessary resources are available is sometimes a distinction between a phase-review process with a forward-looking emphasis on what is needed for the upcoming phase, and stage-gate processes with a retrospective emphasis on stage exit criteria. In order to receive approval for the subsequent phase of a project, the team leader must identify the resources necessary and demonstrate that these are available or could be acquired. Many companies have significantly improved their time to market from these resource-based phase reviews alone.

Yet the confirmation of resource availability is often informal and approximate. Joe *thought* he would have the time available to do the work. Mary *expected* that she would be available in time to do what was needed. John was committed to support the project, but he had made this commitment to other projects as well. While this informal confirmation was much better than nothing, it was still somewhat unreliable.

By incorporating fact-based resource information into its phase-review decision-making process, a product approval committee (PAC) or similar executive team can now make more reliable resource assignment decisions. When a project comes up for approval, the PAC can review the tentative resource assignment profiles if it decides to approve the project. The PAC can see the project assignments and clearly understand potential conflicts. And in preparation for PAC review, the project manager can now work through potential conflicts and resource shortfalls in advance, using resource-based project planning tools.

In addition to avoiding overassigning resources, a PAC can now sequence product development projects to better use capacity. Eventually, this leads to a development rhythm, such as always having one project in the testing phase and offsetting overlapping projects in the development phase in order to maximize capacity utilization. We'll explore this in more detail in Chapter 12.

Resource-Based Project Planning

Some significant improvements in capacity utilization will come by integrating resource management into project planning. Currently, without accurate visibility of resource availability, project team leaders have no choice but to assume infinite availability of resources, even though everyone knows that this is a totally erroneous assumption. Project managers put together their project plans and determine the resources needed. Then they try to find the resources, and when they have a sufficient level of resources committed, they request that their projects be approved. Project managers can only round up these resource commitments one project phase at a time, and when resources are not available for an upcoming phase or phases, they try to manage around the delays.

With resource-based project planning, the team leader can develop a project plan based on actual—or at least close to actual—resource availability. She can adjust the timing of project steps and even major tasks to accommodate the timing with which critical resources will become available. Essentially, she is optimizing the project plan against fixed constraints.

Because of the emphasis on project time to market in the TTM Generation, one of the simplifying assumptions was to start all steps of a project as soon as possible in order to keep as much work as possible off the critical path. If this meant taking resources away from a subsequent project, and the project was delayed or canceled as a result, so what? The focus was on time to market for a particular project, without concern about other projects in the portfolio. With resource-based project planning, a team leader can plan a step to start later, when resources are available, or can consider starting it earlier in order to use available capacity. I refer to this as resource requirements planning, and we'll look at this in Chapter 7.

Resource Group Assignment Balancing

Resource utilization can be significantly improved just by giving resource group managers more responsibility and authority for the utilization of their resources. Simply establishing utilization management as a formal new responsibility leads resource managers to take actions to increase utilization.

In the short term, a resource manager can adjust assignments to increase utilization. He can suggest to a project manager that an assigned resource can split time between two projects, can briefly delay the start of a project to get the resource requested, or can even put more resources on a project to accelerate its development. In many cases, these adjustments increase utilization with no adverse impact on a project and can have a favorable impact. If there is a significant impact, these changes may require approval from the executive decision team.

> Resource utilization can be significantly improved just by giving resource group managers more responsibility and authority for the utilization of their resources.

In the intermediate term, a resource manager can anticipate resource demands on his group and make assignments in a way that better serves anticipated needs. He can better juggle assignments among projects by anticipating upcoming resource needs, and, in some cases, can train his people to be prepared with the skills needed for upcoming projects. Finally, he can use upcoming resource needs as the basis for determining the skills required of new hires.

Longer-Term Capacity Planning and Management

Great benefit can also come from a somewhat longer-term capacity-planning horizon. Longer-term capacity planning is particularly important when a major new product or product platform is on the horizon. Without it, here's an example of what typically happens: A company intends to develop a major new product platform next year, with a tentative start date in mid-year. Without planning a year or more in advance for the capacity needed to develop the platform, resources are routinely assigned to other projects throughout the year because they are ready to commence and offer a good return on investment. But when the time comes to begin marshaling resources for the new platform, most of the required resources are now tied up in other development projects. So the company is faced with the need to either delay the start of the new platform, which is critical to its strategy, or cancel other projects already under development, which wastes the investment to date in these projects.

The best practice here is to establish visibility of development capacity, at least at an aggregate level, well in advance, typically over the next few years. Then create a "placeholder" for the new platform and allocate capacity around it as a constraint. To do this, planned projects must be included in your development pipeline, not just approved projects.

Skill Planning

Having the right mix of skills for the portfolio of projects is critical to resource utilization as well. However, instead of just fitting projects into available resources, a company can go one step further and adjust the skill mix of resources to fit the anticipated needs of projects. In the previous generation, resource managers did not have the information they needed to effectively plan these resource requirements. Without formal capacity management, they had no visibility of the upcoming resource needs of projects in the development pipeline, or of the resource needs of future planned projects. So they made their best guess, and it shouldn't be surprising that they all requested more resources so *they* wouldn't be short when resources were needed. With a formal capacity management process, resource planning can be fact-based, using resource requirements from both actual and planned projects. Using this information, a company can shift resources to relieve anticipated constraints.

Resource Charge-Out

Another resource management practice involves using a charge-out mechanism that fully accounts for the resource costs of development projects by having resource groups formally charge projects for resources they use. In addition to providing more accurate cost information, this practice encourages improved resource management. When a resource is assigned to a project, the project is charged for the cost of that resource, and resource groups are expected to fully charge-out (and thus fully expend) all of their resources. Resource groups are expected to run at break-even over a period of time, such as a year.

When projects are formally charged for the development resources they receive, resource utilization becomes a more critical factor to be managed. At lower utilization, resource costs are much higher, and at higher utilization, resource costs are much lower. Resource groups with low utilization will become more expensive, forcing a focus on efforts to improve utilization and reduce costs. In the longer term, this will lead to more consideration of various types of R&D outsourcing, as companies look at outsourcing alternatives to inefficient or underutilized resource groups.

R&D Outsourcing

As capacity management advances, consideration of outsourcing is a natural extension. Some of the reasons justifying highly vertically integrated development using predominantly captive resources go away with the implementation of a development chain management infrastructure. At the same time, an increased emphasis on R&D capacity management will raise the make versus buy question more often, eventually leading to increased outsourcing of some R&D activities. For example, a company that invests $100 million in R&D may find that as much as $50 million of this could be outsourced to reduce costs and increase flexibility.

This shift toward R&D outsourcing is analogous to the shift away from high levels of vertical integration in favor of increased outsourcing that occurred in manufacturing over the past 20 years. As was the case with manufacturing, financial analysis will be the catalyst for increased outsourcing of R&D. Manufacturers found that it was cheaper to outsource a function if low utilization made the function's cost noncompetitive. This will also be the case with R&D capacity utilization management. For example, a company may find it cheaper to outsource a function such as systems testing if it is unable to utilize its in-house test capability effectively. An independent company may charge $80 per hour, while a dedicated in-house system-test department may cost $120 per hour because utilization is low. (A $60 per hour cost with only 50 percent utilization results in an effective cost of $120 per hour.)

SUMMARY

Getting the most from all of its resources is important to any company, but for companies that rely on a constant stream of new products, the productivity of R&D resources is critical. A company that is excellent at resource management using advanced practices and technologies can achieve 30 to 40 percent greater productivity than a competitor with average resource management performance, enabling it to develop 30 to 40 percent more new products for the same investment.

Yet, despite its importance, resource management traditionally has been given insufficient attention. There are several reasons for this. First, as we've already discussed, the emphasis during the past decade has been on getting individual products to market faster. Second, resource management requires coordinated improvement across all of R&D, and it is more difficult to implement comprehensive change than to implement change at the individual project level. Product development processes in most companies are now sufficiently mature to tackle this change, however. Finally, the systems needed for resource management were not available, and the management concepts and processes needed to apply these systems were not yet developed. These systems, concepts, and processes are now available, and they enable a new generation of resource management, which progressive companies are implementing to gain significant benefits and competitive advantage.

The following chapters set out a roadmap for improving resource management in stages. Each stage has a set of benefits and requires new systems, processes, and practices. The roadmap enables a company to implement resource management in stages, achieve clear benefits at each stage, and yet build a long-term resource management capability.

4

CHAPTER

Stages of Maturity in Resource Management

Resource management is the management of a company's *investment* in development resources, primarily its people, in order to achieve the optimum product development output consistent with its strategy. Rudimentary improvements in resource management were made with the advent of phase-based project management in the early 1990s. For the first time at many companies, the upcoming phase of a project was approved only if sufficient resources were available. This forced project managers to find the necessary resources and, in some cases, to rearrange the company's project priorities in order to win approval. Prior to that time, projects frequently were launched simply because someone wanted to start a new project, with little consideration given to resource availability. These extemporaneous projects created blockages in the product development pipeline that delayed everything and caused significant wasted development. Unfortunately, even the rudiments of phase-based project management still elude some companies, and they are still starting projects without the necessary resources.

The benefits from this very informal resource management were quite significant: Time to market was measurably improved, and wasted development was reduced. We refer to this informal management of R&D resources as Stage 0 in a stages-of-maturity model developed by PRTM to help companies understand why they are not achieving top performance in R&D productivity. Like many new management practices not yet made routine, these concepts and processes may seem complex and daunting. The stages model is an aid to understanding. It also serves another purpose: It provides a roadmap for implementing resource management. Each stage has a specific focus, scope, and a clear set of benefits. Each offers a manageable set of improvements, so resource management can be implemented one stage at a time.

Informal resource management has become insufficient, for several reasons. First, it is informal. Basically, project managers use valuable time soliciting the resources they need for the upcoming phase of a project, requesting approval to free-up or hire specific resources where necessary. Second, it is unreliable. Some people may overcommit to multiple projects, and senior management has no visibility of this overlap in its decision making. Third, the old method of resource management is not done in real time. Information on resource status is inconsistent and frequently out of date. It is also incomplete. Not all assignments are tracked, and there is little control over who is working where. Finally, achieving the benefits of increased utilization discussed in the previous chapter requires a more advanced resource management system and process.

But before we introduce the resource management stages model, let's examine some of the different purposes and characteristics of resource management using a three-level framework.

THREE LEVELS OF RESOURCE MANAGEMENT

Resource management can be confusing, because it serves several different, but interrelated, purposes. Systems and practices designed to serve one purpose don't always work for others. Additionally, what one person means by resource management can be vastly different from what another person means, leading to problems of communication. Resource management is complex, and this complexity needs to be mastered to achieve significant results.

Figure 4–1 defines these different purposes as three different levels of resource management, each with its own objectives and unique characteristics. Level 1 describes resource scheduling across all projects, initially as project resource needs, then translating these needs into specific project assignments. The primary objective is increased resource utilization across all projects. Level 2 describes project resource requirements planning, including planning resource estimates for a project and then later translating assignments to the project step level. Here the objective is to improve project-level productivity, but also to achieve some utilization improvement. Level 3 describes workload management, where the primary objective is the productivity of individual developers.

F I G U R E   4–1

Three Levels of Resource Management

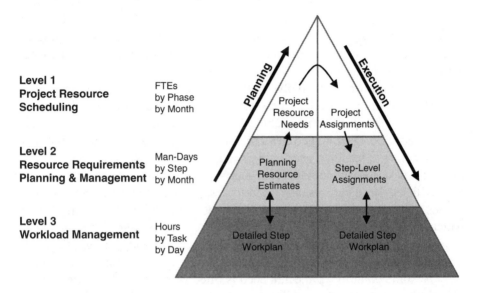

Resource information is defined differently at each of these levels, and the period of time used to manage information is different. The challenge comes in transforming information from level to level for its particular use at each level. It's important to appreciate, too, that resource information always flows in two directions: in the upward direction for planning, and in the downward direction for execution or management.

Each of these three levels of resource management serves different organizational levels. Level 1 serves the enterprise or cross-project organizational level. To be effective, cross-project management must incorporate all resource assignments from all projects. Level 2 resource management serves at the individual project level. An individual project manager can plan and manage her resources at this level. Level 3 serves the individual developer level, or a small group of developers working together on a specific task. At this level, information is applicable to an individual resource or task.

Level 1: Project Resource Scheduling

At Level 1, the objective is resource scheduling and utilization management—keeping track of who is assigned to a project and what resource needs are required. It's at this level that assignments and potential assignments (project needs) are communicated across all projects. Generally, both project resource assignments and project needs are stated as full-time equivalents (FTEs) by month for each phase of a project, such as 0.5 of an FTE for the three months of Phase 1 of the project. This project resource scheduling enables clear communication between project managers and resource managers across all projects.

The data at Level 1 involve both project resource needs and project assignments to fill those needs. These two types of data are sometimes confused or mistakenly used interchangeably. A resource assignment is made when a *specific* individual is assigned to a *specific* project for a *specific* period of time.

In contrast, resource needs are *general*. A need is not based on a specific person; it describes an anticipated resource requirement in general terms, such as a skill set. Needs are usually used to describe planned resources for future phases of a project. Needs and assignments also differ in the direction in which they flow in resource management. A need is used for *planning* resources; it states a potential *demand* for resources. In contrast, an assignment is used for *applying* resources. It states a specific use of resource *capacity*.

At some point in a project, a need is filled with a named resource and becomes an assignment. Generally, this process of requesting needs and filling them, referred to as the resource transaction process, is the primary communication between project managers and resource managers.

Level 1 is the cross-project level of resource management. For it to be effective, all the data on resource needs and assignments across all projects must be captured and maintained. This is the level of resource management at which all project resources are scheduled, and where they are matched to anticipated demand. Essentially, this is the demand and supply balancing process for R&D resources. It is also at this level that demand and supply information is integrated with pipeline and portfolio management to determine where resources should be allocated. We'll get into this relationship in Section Four.

The benefit of improving resource management at Level 1 is a significant increase in R&D utilization, and therefore more output: more new products developed for the same or lower investment in R&D.

Level 2: Resource Requirements Planning and Management

It's at Level 2 that resources are optimized at the project level through more efficient resource requirements *planning* and *management.* This level provides the discipline that helps project managers plan and manage their projects with the fewest resources and the lowest cost while still completing their projects on schedule. Information flows in two directions at this level: *resource requirements planning* to estimate resource needs for a project, and *resource requirements management* to distribute resources assigned across specific project steps.

Resource requirements *planning* enables the project manager to accurately determine resource requirements for the project. It is done at a more exacting unit of measure (person days vs. FTEs) and in more detail than at Level 1, in order to increase planning accuracy. Optimizing project resource requirements involves balancing resource estimates with resource availability for a project, balancing across steps within a phase, and time-phasing steps within a phase. Resource requirements planning can use resources much more efficiently, and, as a result, directly reduce a project's cost and enable a company to complete more projects.

Improved resource requirements planning at this level also has a positive impact on Level 1. Resource requests are apt to be more accurate, and there will be fewer changes to resource requests throughout the project. Since project managers are taking into account the availability of resources in their resource requirements planning, they make much more realistic resource requests.

> An important aspect of resource requirements management is allocating or distributing project resource assignments to specific steps within a phase.

An important aspect of resource requirements *management* is allocating or distributing project resource assignments to specific steps within a phase. This step-specific distribution gives project managers the opportunity to more specifically communicate how the actual resource assignments should be deployed, and enables them to track progress and time against each step. The benefits are increased project control and more projects completed on schedule.

Both resource requirements planning and resource requirements management are explained in Chapter 7.

Level 3: Workload Management

Short-term work scheduling, generally by detailed task, is the purpose of Level 3. Work scheduling is usually done by estimating the hours of work necessary to complete specific tasks, and then summarizing these estimates by day or week to determine the resulting workload. Although workload estimates are very different from resource requirements, these two are also sometimes confused. Workload estimates are intended for and are most useful in the short

term. They help a developer determine what he is going to do over the next
few weeks, and guide the allocation of his time to best accomplish the work.
Workload estimates are much too detailed to be used directly as resource requirements, and task-level details are impossible to plan for in future phases of
a project.

As you might suppose, resource management is best done as an integrated top-down/bottom-up process. Project needs and assignments are

> **Resource management is best done as an integrated top-down/ bottom-up process. Project needs and assignments are planned and managed from the top down, and detailed work estimates are made from the bottom up.**

planned and managed from the top down, and
detailed work estimates are made from the bottom up. The two are then reconciled between
Levels 2 and 3. Resource management done
solely from the bottom up can waste time and
unnecessarily lengthen time to market. We'll
discuss this in more detail in Chapter 7.

Flow between Levels

As you can see from Figure 4–1, there is a continuous flow of resource management information, and as it flows, it is transformed. This is an
important point to remember, because different
layers of the organization use information differently. The person-day resource estimates calculated at Level 2 (Resource Requirements
Planning) flow upward to calculate FTE needs from the management perspective at Level 1 (Project Resource Scheduling). These FTE needs are then translated into assignments, which are then distributed to assigned person-days for
individual steps at Level 2 to manage execution. Detailed workload estimates
from Level 3 are sometimes used to create planning resource estimates at
Level 2, although this tends to be done only when necessary. Finally, detailed
workload estimates can be made for individuals at Level 3 and reconciled to
step-level assignments at Level 2. These resource management levels are best
implemented in stages, as we mentioned, for one stage builds upon the other.
This will all be clearer as we go through this section.

RESOURCE MANAGEMENT STAGES OF MATURITY

Implementing all the resource management capabilities described in this
framework is quite daunting, and a roadmap of some sort is necessary to determine the appropriate sequence for implementing these improvements. This is

especially true for someone implementing a fully capable development chain management system. We'll use a stages-of-maturity model to understand the implementation sequence, and to describe the characteristics and benefits of resource management at each stage. By specifically identifying its current stage of maturity in resource management, a company can also compare its performance in that respect to other companies' performance.

Each stage of maturity represents a progression of the resource management process, each with different objectives and specific benefits. Generally, the stages build upon one another, so a company needs to complete each stage before tackling the next, although some aspects of later stages can be applied in earlier stages. We'll also use this stages framework to describe various aspects of resource management and how they fit together. As shown in Figure 4–2, there are five stages of resource management maturity. The informal resource management described earlier is considered the starting point—Stage 0 of the maturity model. The focus of Stages 1 and 2 is on improving the overall utilization of R&D by getting more productivity from R&D resources, while the focus of Stage 3 is project productivity—helping the project manager to use resources more efficiently. At Stage 4, companies are able to make resource management more seamless and efficient by integrating resource management processes with other management systems and processes.

F I G U R E   4–2

Resource Management Stages of Maturity

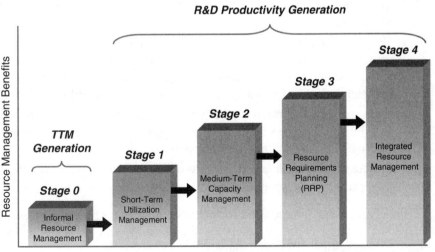

Stage 0: Informal Resource Management

Prior to PACE and similar phase-based decision processes, development projects typically were started without knowing whether the necessary resources were available. Project managers simply assumed that they could find resources when they needed them. The result was a glut of projects that were continually delayed due to lack of resources.

PACE introduced the practice of lining up the necessary resources for the next phase of a project's development as a precondition to receiving approval to commence the phase. This initial attempt at resource management yielded some significant benefits: Projects became more predictable, time to market lessened, and low-priority projects were canceled earlier or never approved in the first place. Eventually, though, companies at <u>Stage 0 resource management</u> capability realized the limitations of this informal approach:

- ◆ Despite efforts to track project assignments through spreadsheets and such, there is no accurate record of who is assigned to what projects, when they are available, and who is not assigned to development projects.

- ◆ Overassigning happens when developers optimistically agree to support more projects than they can, when a change in one project's schedule affects another, or when a project's resource needs change.

- ◆ There is no visibility of resource utilization. The vice-president of R&D cannot answer the question, "What percentage of our resources is working on approved projects?" As expected, development capacity utilization tends to be much lower in the absence of visibility.

- ◆ Resource managers do not have a record of the project assignments of developers in their groups because it is too time-consuming to gather the necessary information. Frequently, they must resort to asking each individual about his or her expected availability.

- ◆ With any informal process, there are inconsistent understandings of assignments between the project managers and resource managers, leading to frustration and disruption.

- ◆ Project managers have no visibility of resource availability, so they plan projects assuming infinite availability.

Finally, in Stage 0 there is no formal system for managing resources. The management process is based solely on relationships and practices between project managers and resource managers.

Stage 1: Short-Term Utilization Management

Companies at a Stage 1 level of maturity in resource management at least have a process for project-assignment and resource-utilization management. Formal resource assignments are made to all projects, and resource assignments are managed for all resources. Three resource management practices are in place at Stage 1: (1) project resource assignment, (2) management of assignments by resource groups, and (3) utilization management based on assignments.

Moving from Stage 0 to Stage 1 provides immediate utilization benefits. Formally measuring utilization gives a company insight into the projects that its R&D resources are actually working on, and generally there are some obvious actions it can take to quickly improve utilization. As we'll see in the next chapter, these quick benefits can be quite significant. Stage 1 capability also provides immediate benefits to project managers and resource managers. Project managers have the tools to check resource availability and explicitly keep track of who is working on their projects as well as the costs associated with those resources. Resource managers have visibility of where the people working for them are assigned, as well as a scheduling tool for managing the availability of their people. Utilization reporting based on assignments enables companies, for the first time, to measure the utilization of R&D at various levels and to identify objectives for improvement. As was explained in the previous chapter, the benefit of increased utilization can be quite significant. Stage 1 maturity is also an appropriate point at which to clarify resource costs, since it is now possible to assign the costs associated with resources to projects more accurately than was possible in the past.

> Utilization reporting based on assignments enables companies, for the first time, to measure the utilization of R&D at various levels and to identify objectives for improvement.

Stage 1 resource management is a good starting point because it is simpler than Stage 2, since at Stage 1 a company is simply formally keeping track of all project assignments, which all companies are already doing informally. At Stage 1, companies don't have to define project needs or set up a formal resource-transaction process. But while Stage 1 confers quick benefits, those benefits are generally not sustainable without moving on to Stage 2. Please note, however, that the benefits of Stage 2 are not attainable without a mastery of Stage 1 practices.

Stage 2: Medium-Term Capacity Planning and Management

At the completion of Stage 1, a company has visibility of *all* project assignments, it is able to measure utilization based on assignments, and its project and resource managers have the direction and tools to improve resource utilization in the short term. But once Stage 1 is reached, managers become frustrated. They're frustrated that their visibility of resource needs does not extend beyond assignments, that they don't have enough information to do any capacity planning, and that they don't have an automated system for managing resource transactions. Companies can only address these frustrations by moving to a Stage 2 level of maturity.

Companies at Stage 2 are able to do medium-term resource capacity planning and management, to look further ahead at anticipated resource needs, and to make adjustments to resource capacity and resource demand in order to increase output. Essentially, Stage 2 entails the supply-demand balancing aspect of resource management. Stage 2 companies have development chain management systems that enable them to determine R&D budgets based on realistic capacity needs, and to help them extend their visibility into the period covered by their annual financial plans.

It's at Stage 2 that companies begin to understand resource needs and gain the necessary visibility for medium-term capacity resource management.

Companies at Stage 2 are also able to address the management of the resource transaction process. This is the process that controls the initiation of resource requests from needs identified by project managers, the response to these requests with assignments by resource managers, and the coordination of all the exceptions, changes, and adjustments to these resource transactions. In a large company, hundreds of resource transactions take place every week, and an effective system and management process can make this otherwise cumbersome and unreliable process much more efficient and reliable.

It's at Stage 2 that companies begin to understand resource needs and gain the necessary visibility for medium-term capacity resource management. Needs are translated into assignments when a project manager makes a resource request from a resource manager, with the resource transaction process managing the translation. While some companies find it best to implement resource capacity planning and the resource-transaction management process at the same time, others find it easier to do capacity planning first and then implement the resource transaction process later. To accommodate the latter, the introduction of resource needs and resource capacity planning is sometimes referred to as Stage 2A, and the resource transaction process as Stage 2B.

Stage 3: Resource Requirements Planning (RRP)

Companies at a Stage 3 level of capability have extended their focus beyond those at Stage 1 or 2. In those stages, the focus is on resource management across *all* projects. Assignments and needs are collected at a high level (Level 1) from all projects and used to manage all resources. The objective is optimization of resources *across* all projects (short-term in Stage 1, and medium-term in Stage 2); there is no concern about the efficiency of resources on any project.

In Stage 3, the objective is to optimize resources for *each* individual project. Resources are estimated, planned, and managed at the step level, enabling project managers to optimize the use of resources on projects and reduce project costs.

The benefits in this stage can be quite significant and strategically important. For example, if each project can be completed with 20 percent fewer resources, then the return on investment will increase dramatically. Take a project with a $6 million development cost, including a $5 million cost for developers, and an expected $9 million profit contribution (50 percent ROI above the development cost). Through improved resource requirements planning and management, the resource cost can be reduced to $4 million and the project cost to $5 million, while still completing the project in the same time. As a result, the project ROI improves to 80 percent [$9M – $5M / $5M]. Companies at Stage 3 capability enjoy project portfolio benefits as well. A 20 percent lower resource requirement across all projects means that 20 percent more projects can be completed with the same total resources, so overall output increases by 20 percent.

Stage 3 resource management capability has three major elements. *Project resource requirements planning*—the planning of Level 2—enables project managers to effectively plan resource requirements. *Project resource requirements management,* also a Level 2 capability, enables project managers to effectively manage resource assignments. *Workload reconciliation,* enables project team members to reconcile detailed work-scheduling estimates with their project step assignments.

Stage 4: Integrated Resource Management

With the basics of multiproject assignment-based utilization management and project-centric resource optimization in place, companies can proceed to Stage 4: the extension and "institutionalization" of their resource management improvements and benefits. At the Stage 4 level of capability, companies focus on integrating other systems with the resource management system, and other processes with the resource management process. The primary benefits of this stage are the seamless operation of a company's systems and processes and the increased efficiency it brings.

> At the Stage 4 level of capability, companies focus on integrating other systems with the resource management system, and other processes with the resource management process.

Resource management integration can be quite broad. In the chapter discussing this stage, we will cover several types of integration. Resource management can be fully integrated with project budgets so that when the project manager changes assignments or needs, the project budget automatically changes. The integration of resource management with the system and process for product strategy enables managers to automatically include planned products in their resource planning. This capability allows resource management to be integrated with the annual planning process so that R&D budgets can be based on realistic resource requirements.

Another typical form of integration enabled at Stage 4 is closed-loop time collection, whereby R&D time worked is collected based on scheduled time and fed back directly to the project manager. Integration of R&D resource management with external resources, such as codevelopment partners and other outsourced partners, can also be accomplished at this stage. A company can then integrate the management of external resources into its overall resource management process and system.

How the Levels and Stages Work Together

At this point, it's helpful to review how the stages of resource management were derived from the three-level resource management framework that we discussed at the outset. Remember that the purpose of the stages model is to define a reasonable roadmap for implementing improvements, and that sections of the three-level framework were identified and prioritized to accomplish this (see Figure 4–3).

Stage 1 addresses project resource assignments, and provides very significant benefits from increased utilization at the start and initiates resource management. Stage 2A then builds on Stage 1 by adding project resource needs to the resource management process, leading to further utilization benefits. While Stage 1 is focused on the short term, Stage 2A is focused on the medium term.

F I G U R E   4–3

Three Levels of Resource Management and Stages of Resource Management

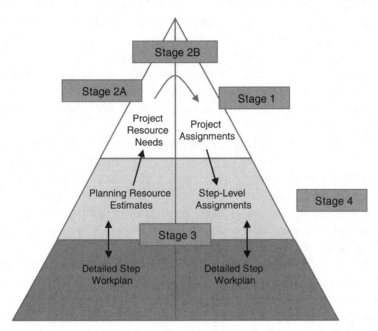

Stage 2B adds the resource management transaction process to link Stage 1 and Stage 2A, completing the Level 1 resource management process of project resource scheduling.

Stage 3 covers Level 2, and its integration with Level 3. Since Level 2 can be implemented project by project, and Level 3 can be implemented by individual developer or project step, the implementation of Stage 3 can be rolled out over time, and a partial implementation can be partially successful. Stage 3 thus covers a lot more of the three-level framework. Stage 4 is pictured externally to the three-level framework but it really involves the integration of this framework with other systems and processes. Here again, this will become clearer as we progress through this section.

ROADMAP FOR IMPROVING RESOURCE MANAGEMENT

When a company understands the stages model of resource management maturity, it can establish its current baseline and then develop a roadmap for making progress toward targeted goals.

Our hypothetical company, Commercial Robotics Inc., had always expected the highest performance in all areas of the company, and project resource management was no exception. Brian Kennedy, the CEO, challenged the VP of R&D, Chris Taylor, to improve R&D productivity. "We have no limit to the opportunities to develop new products; we're only restricted by how much we can invest in R&D without reducing our profits," he said. "We can't spend any more on R&D without reducing our profits and hurting our stock price. We understand how good this investment is, but, unfortunately, the stock market doesn't. So what I need to know is, how can we get more output from the same input?"

Taylor decided to use the resource management stages model to determine CRI's baseline performance and to develop a roadmap for improvement. It was immediately obvious that CRI was at Stage 0. "We have tried several times to get control over our resources, but without much success," Taylor conceded. "We generally used spreadsheets or other simple tools to track assignments, but this wasn't the right approach. We are now serious about this. This is a priority for us, and we are going to do it right."

In the next four chapters, we will follow CRI as it progresses through the four stages of increasing maturity in resource management. Its story will illustrate the capabilities and benefits that advanced processes and systems can enable.

SUMMARY

In this chapter we reviewed two underlying concepts for resource management. The three-level framework defines resource management according to its different purposes. The stages of resource management framework defines a progression of resource management capabilities that enables specific improvements, and more importantly, specific benefits.

5

CHAPTER

Stage 1—Short-Term Utilization Management

Stage 1 is generally the starting point for most companies that are implementing resource management systems and processes because it is the easiest first step, it provides the necessary foundation for future stages, and, most importantly, it creates some significant early benefits. Short-term utilization management is the primary objective of Stage 1, and managing project assignments is its basic focus.

Three primary resource management practices are introduced at a Stage 1 level of maturity, and these are contingent on the use of a common resource management database (Figure 5–1), which is part of a DCM system. Project managers keep track of resources assigned to their projects, resource group managers keep track of the assignments of all of their people, and utilization is reported based on assignments. The key to Stage 1 is that all three of the aforementioned activities are based on common resource assignment data.

Without this common database, most companies informally manage assignments, and assignment information is neither centralized nor consistent. Project team managers keep track of assignments using their own spreadsheets, emails, notes, or their personal recollections. Resource managers informally keep track of the assignments for resources in their groups in similar ways. Project and resource management data are frequently inconsistent; there

F I G U R E   5–1

Three Primary Stage 1 Resource Management Functions

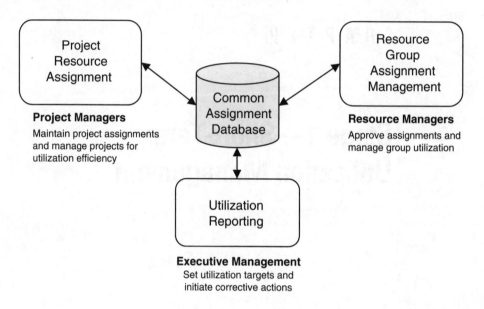

Project Managers
Maintain project assignments
and manage projects for
utilization efficiency

Resource Managers
Approve assignments and
manage group utilization

Executive Management
Set utilization targets and
initiate corrective actions

is no official record of resource assignments, so accurate utilization reporting is impossible.

Once resource assignments are formally managed within a company, project managers know exactly what resource assignments have been made for their projects, resource managers know where all their people are working, and the company has an accurate view of current utilization across all resources. There are both long-term and short-term benefits from these capabilities. We'll be focusing on the short-term benefits in this chapter.

RESOURCE ASSIGNMENT

With a formal resource assignment system as part of an integrated DCM system, all project managers use the same system for project assignments, and their assignments update a common assignment database. Approval from resource managers to assign a particular developer to a project may still be informal. The process of making assignments may be as simple as discussing assignments with resource managers, holding joint planning sessions where everyone is in the same conference room, or through other ways of communicating, such as email. The addition of a formal resource transaction process to manage this process is implemented in the next stage.

At this stage, the question typically raised is, who assigns resources to a project? The project manager, or the resource manager? While this is a basic question, it's remarkable that in some companies there is still confusion over the answer. It's useful for companies in Stage 1 to answer the question more clearly, and although informal practices do vary, it's important to recognize the role of resource managers in project assignments. For companies planning to continue resource management improvements through Stages 2A and 2B, the responsibility for assigning resources to a project in Stage 1 should be defined as a step in that direction. In the remainder of this chapter, we'll treat the details of resource assignment somewhat generically.

An important element of assignment management in companies that use a phase-review or similar project-approval process is the ability to track the approval status of assignments. Resource assignments to an approved project phase are "approved" assignments and generally have a different status than "planned" assignments for future project phases yet to be approved.

Project Resource Assignment

Typically, a formal assignment management system provides the project manager with several helpful tools that make resource assignment easier and more effective. In an integrated development chain management system, these tools are integrated with enterprise project planning, which we will look at in Chapter 10. Because there is a common assignment database, the project manager can easily check on resource availability when considering assignments. She can also keep a running total of project cost as assignments are made, enabling her to balance project cost and assignments.

Anne Miller, the Fast-Food Robot project manager at CRI, found these tools and the formal assignment management system CRI had installed to be extremely helpful. Figure 5–2 illustrates the resource assignments to Phase 1 of her project, which she made in conjunction with the appropriate resource managers. Anne assigned herself full time (1.0 full-time equivalent) to the project for the six months of Phase 1. She initially intended to assign Richard Salisbury from Marketing full time as well, but in checking his availability she found that he had some other project commitments from March through May. With visibility of Richard's project assignment load, she was able to determine that he could do the necessary work on Phase 1 of her project while supporting the other projects to which he was assigned as well. She assigned him full time for January and February, half time (0.50) for March through May, and full

FIGURE 5-2

Project Assignments for the Fast-Food Robot Project at CRI

| | | | Phase 1 — Planning | | | | |
	Jan	Feb	Mar	Apr	May	June	Total
Project Management							
Anne Miller	1.00	1.00	1.00	1.00	1.00	1.00	6.00
Marketing							
Richard Salisbury	1.00	1.00	0.50	0.50	0.50	1.00	4.50
Software							
Naoto Takeyama	0.50	0.50	1.00	1.00	1.00	1.00	5.00
David Krieger	0.00	0.00	1.00	1.00	1.00	1.00	4.00
Finance							
Art Hall	0.50	0.50	0.50	0.50	0.50	0.50	3.00
Mary Bailey				1.00	1.00	1.00	3.00
Electrical Engineering							
Alan Tan	0.25	0.25	1.00	1.00	1.00	1.00	4.50
Mechanical Engineering							
Lynda Stevens	0.25	0.25	1.00	1.00	1.00	1.00	4.50
Harold Welch			1.00	1.00	1.00	1.00	4.00
Quality							
Richard Ingram	0.25	0.25	0.25	1.00	1.00	1.00	3.75
Project Total	3.75	3.75	7.25	9.00	9.00	9.50	42.25
Project Cost (000)	$54	$54	$105	$131	$131	$138	$613

time again in June, when he would be needed to help prepare for the Phase 1 Review.

Anne needed two software engineers, and was able to get Naoto Takeyama phased-in to full time after the first two months, during which he was completing another project. She was happy with this since Naoto had worked with her on a similar project, and she really wanted him to be the senior software engineer for this one. She needed another software engineer, and the software engineering manager assigned David Krieger, a new hire, to the project. This was fine, but Anne asked that he not start on the project until Naoto was full time so that he could supervise David from the start.

Anne really wanted Art Hall to do the finance work on the project, since he was the most senior financial manager and this was her personal weakness, but he was only available half time, and more time than this would be necessary to complete the work in Phase 1. Anne got around this by adding a junior financial analyst, Mary Bailey, to help Art from April through May, when most of the number-crunching was anticipated. Art liked this a lot better, and to Anne's surprise this actually cost her project less than having Art full time. Anne then completed her project team with Alan Tan from Electrical Engineering, Lynda Stevens and Harold Welch from Mechanical Engineering, and Richard Ingram from Quality. Instead of simply requesting all of these people full time, she ramped up their time assignments to her project based on how much they would be involved in Phase 1 and when they would be available.

Anne reflected on how this new assignment management system changed the way she made project assignments. "I approached this very differently. In the past I would have simply identified the people I wanted, said I had to have them full time, and then got ready to fight for my project as the most important one. I'm actually pretty good at that, and usually win, but it's a frustrating process. The new system has motivated me to be more flexible and try to balance a developer's availability with the real needs of the project. I've also made more part-time assignments than I would have in the past. Previously, if you didn't take someone full time, and take them immediately, then you lost them, but now I have more confidence that, with common assignment information, when I reserve someone's time it will be respected, and overassignments will be minimized."

When asked if there were any other benefits, Anne replied, "Yes, it cost my project less. For kicks I figured out what the resource cost would have been if I had made assignments the old way, and it was almost $100,000 more because I would have traditionally made more full-time assignments. That's almost a 15 percent savings. In fact, in the past, I really didn't think of the resource cost until after I made the assignments. Now I can see in real time the impact of assignments on project cost, and it makes me much more conscious of the cost of alternatives."

With these assignments, the resource cost for Phase 1 of this project is estimated at 42.25 person months, with a cost of $613,000. Anne has some estimates of resource costs beyond this phase, but these estimated costs are not yet explicitly defined as anticipated resource needs beyond these assignments This capability is not possible until a company adopts the practices and systems of Stage 2.

Resource Group Assignment Management

With a shared assignment management system in place, enabling formal assignments that are visible to all managers, resource managers can see, for the first time, where their resources are assigned and share formal agreements with project managers about how those resources will be used.

One of the resource managers at CRI talked about the way he kept track of his resources before the company implemented this common assignment system. "Previously, when one of my people was assigned to a project, the project manager assumed that he owned the person until he gave him back, and most of the time I really had no idea of when that would be. When another project manager wanted a resource, I had to call around to see when my own people might be available. To get more control, I started to confirm all assignments with emails, and would use them later to verify that my understanding was correct, but this only worked some of the time and usually incited disagreements.

With a shared assignment management system in place, enabling formal assignments that are visible to all managers, resource managers can see, for the first time, where their resources are assigned and share formal agreements with project managers about how those resources will be used.

"Formal assignment management greatly improved this. Now I can see exactly where my people are working and when they are available. In fact, everyone in the company can see their availability, and make more intelligent requests based on availability. In addition, I can assign my people to new projects based on when they are scheduled to complete the ones they're currently on. Most importantly to me, project managers no longer think that they own my people until they give them back."

Within an assignment management system, resource assignments are combined with other items scheduled by resource managers, such as vacations, training, and internal projects, in order to provide a complete view of resource availability. Once the resource manager has this visibility, he can make tactical adjustments to improve utilization of the resources he manages.

Figure 5–3 demonstrates how this complete picture of a project's resource utilization was created by Ted Johnson, manager of the Software Engineering Resource Group at CRI. As can be seen by the totals at the bottom of the table, Ted currently has 25 software engineers in his group, and the group has 88 percent utilization in January. As shown, 22 people are assigned to approved projects, 2.5 have other assignments, and 0.5 are available. The availability percentage increases every month, up to 32 percent in April, but this is to be expected, since this utilization is based only on assignment data.

As a result of this resource management capability, Ted was able to make adjustments to improve utilization and better meet project resource

F I G U R E 5–3

Software Engineering Group Resource Manager's View of Assignments at CRI

Resource Group: Software Engineering

Resource	Assignment	Jan	Feb	Mar	Apr	May	June	July	Aug
Naoto Takeyama	Projects:								
	Fast-Food Robot	0.50	0.50	1.00	1.00	1.00	1.00		
	Monitoring System	0.50	1.00						
	Nonproject:								
	Training								
	Vacation		0.25						
	Total	1.00	1.75	1.00	1.00	1.00	1.00	0.00	0.00
	Available (Overassigned)	0.00	-0.75	0.00	0.00	0.00	0.00	1.00	1.00
David Krieger	Projects:								
	Fast-Food Robot			1.00	1.00	1.00	1.00		
	Industrial Software V10	0.50	0.50						
	Nonproject:								
	Training							1.00	
	Server Upgrade	0.50							
	Total	1.00	0.50	1.00	1.00	1.00	1.00	1.00	0.00
	Available (Overassigned)	0.00	0.50	0.00	0.00	0.00	0.00	0.00	1.00
Jennifer Jones	Projects:								
	Industrial Robot V2			1.00	1.00	1.00	1.00	1.00	1.00
	Industrial Robot Special	1.00	1.00	1.00					
	Nonproject:								
	Training							1.00	
	Vacation								0.50
	Total	1.00	1.00	2.00	1.00	1.00	1.00	2.00	1.50
	Available (Overassigned)	0.00	0.00	-1.00	0.00	0.00	0.00	-1.00	-0.50
Harry Oh	Projects:								
	Industrial Robot V2			1.00	1.00	1.00	1.00	1.00	1.00
	Industrial Cleaner	1.00	1.00	0.50					
	Nonproject:								
	Training							1.00	
	Customer Support	0.50	0.50						
	Total	1.50	1.50	1.50	1.00	1.00	1.00	2.00	1.00
	Available (Overassigned)	-0.50	-0.50	-0.50	0.00	0.00	0.00	-1.00	0.00
Susan Collins	Projects:								
	Industrial Cleaner	0.50	0.50						
	Industrial Software V10	0.50							
	Nonproject:								
	Training							1.00	
	Vacation								0.50
	Total	1.00	0.50	0.00	0.00	0.00	0.00	1.00	0.50
	Available (Overassigned)	0.00	0.50	1.00	1.00	1.00	1.00	0.00	0.50
Total	Scheduled on Projects	22.0	20.0	18.0	15.0	15.0	16.0	15.0	10.0
	Percent Utilization	88%	80%	72%	60%	60%	64%	60%	40%
	Other	2.5	3.0	2.0	2.0	2.0	2.5	5.0	10.0
	Total	24.5	23	20	17	17	18.5	20	20
	Available	0.5	2	5	8	8	6.5	5	5
	Percent Available	2%	8%	20%	32%	32%	26%	20%	20%

needs. "Naoto is currently overassigned in February at 1.75, so I talked with him about delaying the vacation he originally scheduled for February. I also talked with the project manager of the Monitoring System project to see if he could continue at half time instead of going back to full time as was originally planned, and this looks feasible."

"David Krieger has half-time availability next month. I'm going to see if he can start a little earlier on the Fast-Food Robot project, or if his available time can be used on the Industrial Software V10 project." Remember that earlier we saw how Anne Miller intentionally delayed David's start on this project, so Anne and Ted will need to work this out.

Ted is also aware of a second overassignment. "Jennifer Jones is double-scheduled in March; she is assigned full time on both the Industrial Robot V2 project and the Industrial Robot Special. I'm not sure how that happened, but I suspect that the Industrial Robot Special is taking longer, and the project manager extended her on the project without asking me. I'll talk with him about it, and I expect that with a two-month visibility we can work something out. Previously, I didn't know about issues like this until it was too late to make any constructive adjustments. I also need to talk to the project manager about July and August, when Jennifer is scheduled for vacation and training. We'll have to get someone else lined up then."

"Harry Oh is overscheduled in January and February, because he is assigned to customer support. I can have someone else in my group cover customer support instead. Then I need to look at Harry's time in March, where he is overscheduled on two projects, but I expect that he can phase into the Industrial Robot V2 project. Finally, I need to discuss his training too."

"Susan Collins is a new software engineer with us. She is available in February, and I need to find a project for her. I'll talk with a couple of project managers and see if they could use some additional help. If that doesn't work out, then I'll put her on customer support and schedule some additional training for her."

Ted Johnson concluded by sharing his objectives for resource management. "I want to keep my group at 85 percent utilization. Eighty percent is the standard, but I think I can do better than that. It's more difficult in the summer months, when there are a lot of vacations, so I try to run at close to 90 percent in the other months." Ted also commented on the new resource management tools. "Without this visibility I couldn't possibly manage resources. I expect that making adjustments, such as those I discussed, will enable me to increase utilization by 5 percent, maybe more. This 5 percent increases productivity by about one software engineer, or approximately $150,000 per year, and it's now a big part of my performance evaluation."

At this stage, the resource manager has actual assignment information, but not visibility of anticipated resource needs beyond immediate assignments. When this visibility becomes available, intermediate and longer-term resource decisions can be made. But that capability will require more advanced systems capabilities that we will discuss in the next chapter.

Utilization Reporting

With a common and accurate project assignment management system, a company not only has the information needed for project managers and resource managers, but also has accurate information for managing utilization. It can measure the percentage of resources used on approved product development projects and can identify what everyone else is doing. Some of the reports that can be generated are illustrated in Chapter 3. Figure 3–3 illustrates a utilization trend report for all of the R&D resources at CRI, and Figure 3–4 illustrates a utilization report for the Software Engineering Group.

Utilization reports based on assignments essentially take a snapshot of where people are assigned at a particular point in time—usually at the end of the month. A series of these snapshots enable trend reporting, showing how utilization is changing from month to month. Utilization reports can also compare actual utilization to standard or planned utilization levels.

As described in Chapter 3, utilization is an important measure of R&D productivity and the primary metric for quantitatively measuring the improvements of Stages 1 and 2, so it's important to establish a utilization baseline and to measure improvement from the baseline. As illustrated in Figure 3–3, CRI estimated that the 10 percent improvement in utilization equated to $6.8 million in cost savings or bottom-line gain.

Utilization management involves more than just utilization reporting. It also requires a management process, which starts by defining responsibilities for utilization. What are the responsibilities of the R&D vice-president, resource managers, project managers, and individual developers? They all have some responsibility, since they all have some control over utilization, yet at the same time there is a certain level of interdependence that must be recognized. A clear definition of responsibilities must be set out up front before each of these functions can be held accountable. At this point it's also important to train everyone on how he or she can influence utilization and productivity, and clearly state what is expected. Executives should clearly establish performance expectations, decide if these will be increased over time, and integrate these expectations into performance appraisals where appropriate.

RESOURCE ASSIGNMENT COSTS

Formally assigning resources to projects usually raises the related issue of
more formally assigning the costs of those resources to projects as well. At
this point, now that a company has accurate assignment information, it's appropriate for it to reconsider how it computes resource costs.

Resource costs are frequently underestimated in evaluating product development projects. In some cases, this leads to approving projects that would
otherwise have been canceled, and it encourages project managers to request
more resources than they need in order to cover contingencies that might arise.

It's helpful to view resource costs as multiple cost layers. Figure 5–4 illustrates how CRI computed its resource cost layers. The average annual R&D
salary at the company is $80,000, and benefit costs are approximately 25 per-

F I G U R E 5–4

CRI's Estimated Annual Resource Cost

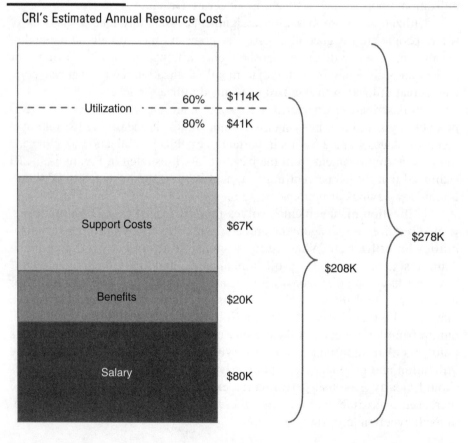

cent of salary, or another $20,000. (These include medical insurance, the company's share of payroll taxes, unemployment taxes, etc.) On top of this, there is another $67,000 per person of what CRI refers to as support costs. These include facilities, training, computer and laboratory equipment, supplies, telephone, and nonproject travel. Other costs directly incurred for a project, such as project travel, are charged directly to the project. This brings resource costs to $167,000 per year per developer.

CRI recognizes that these costs are implicitly affected by utilization. If a developer is 80 percent utilized, then his $167,000 cost spread over 80 percent of his time equates to $208,000, and if he is only 60 percent utilized, then his effective cost equates to $278,000. Even though CRI had only 60 percent utilization, it decided to use the cost that equated to $208,000 because it was committed to getting utilization up to 80 percent.

> Resource costs are frequently underestimated in evaluating product development projects.

Defining formal resource costs is an important element in Stage 1 resource management because it emphasizes heightened accountability for resource assignments. This accountability emphasizes how valuable development resources really are. With full accounting for resource costs, a company will finally understand the real costs of its R&D.

Take, for example, a company that charges resources to projects at straight salary cost. In CRI's case, only approximately 30 percent [$80K / $278K] of its R&D cost would be charged to projects, leading to significantly understated product development costs. It's likely that many projects that would not have been approved under more realistic resource-cost estimates would be approved under this salary-based charging scenario, with its artificially low cost assumptions and therefore unrealistically high ROI estimates. This may explain why some companies have a poor return on overall R&D investment even though each individual project *appears* to be justified on an ROI basis.

REQUIREMENTS

Companies operating at a Stage 1 level of resource management must maintain all project assignments for all R&D projects. If they don't, incomplete and inaccurate assignment data will make it impossible to distinguish who is available from who is assigned but not recorded as assigned in the database. To achieve Stage 1 capability, then, companies must have an appropriate resource management system, processes, and policies for managing assignments.

They must also clarify organizational responsibilities for resource management and utilization.

Stage 1 management capability requires a development chain management system for managing project assignments. It simply cannot be done with spreadsheets or project plans distributed across the computers of project managers and resource managers, because these spreadsheets and plans are not coordinated. Thus, what the project manager records may be different from what the resource manager records. When resource assignment data are distributed, project managers and resource managers each need to post the same data and manually synchronize any changes, and, as a result, there are no common assignment data for utilization reporting. The use of shared spreadsheets can have some favorable impact, but eventually becomes inaccurate and inefficient. Spreadsheet sharing still amounts to trying to use a tool when a system is required.

> **The use of shared spreadsheets can have some favorable impact, but eventually becomes inaccurate and inefficient. Spreadsheet sharing still amounts to trying to use a tool when a system is required.**

A resource assignment management system is also very different from detailed project planning tools that record and summarize work estimates. As was explained in Chapter 4, resource assignments are very different from detailed work estimates. They are made at a higher level and are defined by a time period, not a task. Detailed project workplans also tend to be inconsistent and unreliable for purposes of consolidating resource assignment information. The assignment management system must maintain a central assignment database that both project managers and resource managers use as their exclusive source of assignment information. This assignment database is also the source for utilization reporting.

If a company uses a phase-review or stage-gate process, then it's also critical that the resource management system supports assignment by phases. Otherwise, it will undermine the integrity of the phase-review process by implying that resources assigned to future phases are approved, even though these phases have yet to be approved.

Prior to selecting a project resource-assignment management system, a company should determine if it intends to continue improving resource management to a Stage 2 capability. If it does, then it won't want to completely replace its resource management system. Therefore, the initial resource management system must be used to manage resource needs, which are introduced in Stage 2, as well as resource assignments, since assignments are based on needs.

Effective Stage 1 resource management requires more than just implementing a project assignment management system; it also requires the necessary management processes and policies. What are these processes and policies? A more formal resource assignment management process defines the following:

- When project assignments are made, and the duration of assignments (best practice is that assignments are made by phase, not for the duration of the project)
- The format for assignments (person days, FTEs, percent of time, etc.) and consistent practices for partial assignments
- The procedure for requesting assignments from resource managers
- The procedure for consistently planning nonproject time
- The process for determining resource costs (by person, by level, by group, and extent of costs)
- The definition and format for utilization reporting
- Utilization targets

Implementing the more formal practices of Stage 1 resource management requires clearly defining responsibilities and authorities that may have previously been understood informally or intuitively. While loose understandings might suffice for a few projects and a small staff of developers, they will not suffice for a larger-scale operation.

Who makes the decision to assign someone to a project? What are the responsibilities of project managers in requesting assignments? Who actually enters the assignment in the database? Who is responsible for the accuracy of assignments? Generally, in Stage 1 resource management, project managers directly make project assignments after reaching agreements with resource managers. In some cases, a company may have someone who acts as a project resource administrator to maintain all project resource assignments, especially in the early implementation of Stage 1.

It's also necessary to clarify utilization responsibilities. Are resource managers measured and evaluated on the utilization of their resources? What are the responsibilities of project managers? Are the R&D VP and other executives with development staff under their control evaluated on their resource utilization? Part of this evaluation involves establishing utilization standards, and these need to be set appropriately.

BENEFITS

The benefits from Stage 1 short-term utilization management are quite significant. These were described throughout this chapter, but it is useful to summarize them at this point:

1. The initial utilization baseline, if accurately determined, will almost
   always identify the opportunity for some significant utilization improvements. It will uncover resources that are overscheduled *before*
   they delay projects. The baseline will generally show that utilization
   is lower than expected, and will identify the various activities other
   than approved development projects on which R&D resources are
   working. From this utilization baseline, R&D management can generally take some quick actions to reassign developers and increase
   utilization.

2. Utilization reporting provides the basis for continuous improvement
   and establishes utilization as an important objective. This will
   change the behavior of resource managers and project managers to
   further increase utilization.

3. With the right tools and resource assignment management practices
   as part of an integrated resource management system, resource managers can easily make adjustments to assignments and increase utilization. And project managers can plan projects to more effectively
   use resources, in many cases reducing resource assignments and
   therefore project costs.

4. By linking resource assignments directly to project costs, and by
   more formally assigning resource costs, project managers will be
   more efficient in using resources, and projects will be more appropriately evaluated, leading to greater availability of resources for
   other projects.

5. Assignment information only needs to be recorded once, saving
   unnecessary administrative work. This one-time entry also eliminates much of the friction between resource managers and project
   managers.

6. Overassignments are avoided, since availability is visible to everyone. This, in turn, will eliminate many project delays.

Overall, companies at a Stage 1 level of resource management capability
can quickly improve utilization by at least 10 to 20 percent, depending on the
company and its success in implementing the various Stage 1 systems and

practices. Utilization improvements translate directly into increased R&D productivity. For a company spending $100 million on R&D, these improvements would equate to a $10 million to $20 million savings per year. Improved utilization also reduces the cost of individual projects, as project managers become more careful with their resource assignments.

When a company completes its Stage 1 resource management improvements, it will reap significant, measurable benefits, but it will also see more clearly what gains it will achieve by implementing Stage 2 capabilities. These are the subject of the next chapter.

SUMMARY

In this chapter, we introduced Stage 1 resource management by defining its primary objective: short-term utilization improvement. Stage 1 resource management encompasses three major practices—project resource assignment management, resource group assignment management, and utilization reporting—all around a common assignment database. We discussed each of these practices in some detail, and illustrated how they were applied by our hypothetical company, CRI. We also reviewed the concepts of accurately computing resource costs with respect to resource utilization.

6

CHAPTER

Stage 2—Medium-Term Resource Capacity Planning and Management

Companies at a Stage 1 level of capability in resource management have visibility of all project assignments, and are able to measure and manage utilization based on assignments. Project managers and resource managers have the direction and the tools to improve resource utilization. Having accomplished this, some managers are now frustrated that their visibility of resource needs does not extend beyond assignments, that they don't have enough information to do any capacity planning, and that they don't have an automated system for managing resource requests and assignment transactions. Stage 2 resource management capabilities focus on medium-term resource capacity planning and management, providing the ability to look further ahead at anticipated resource needs and to make adjustments to resource capacity and demand in order to increase output.

Companies with Stage 2 capability also are able to address the management of the resource transaction process. This is the process that controls the initiation of resource requests for approved projects from needs identified by project managers, the response to these requests with assignments by resource managers, and the coordination of all the exceptions, changes, and adjustments to these resource transactions. In Stage 2, the responsibility for resource assignments generally shifts decisively to resource managers. In a large company, hundreds of resource transactions take place every week, and an effective system and management process can make this otherwise cumbersome and unreliable process efficient and reliable.

It's at Stage 2 that companies understand the concept of resource needs. Visibility of future project resource needs allows them to do medium-term resource capacity management and supply-demand balancing. Needs are translated into assignments when a project manager makes a request of a resource manager and the resource manager fills the request with an assignment. The resource transaction process manages this translation. While some companies find it best to implement resource capacity planning in conjunction with the resource transaction management process at the same time, others find it easier to do resource capacity planning first and then implement the resource transaction process later. To accommodate these two alternatives, you'll remember, the introduction of resource-needs and medium-term resource capacity planning is sometimes referred to as Stage 2A, and the resource transaction process as Stage 2B.

RESOURCE CAPACITY PLANNING

In understanding resource planning, it's helpful to start by looking at what can be done to plan capacity over different time horizons. Resource planning takes place over three time horizons, and each uses a different source of information and has a different purpose.

Short-term resource planning, which we discussed in the previous chapter, is based on project assignments. Its purpose is to increase utilization and satisfy immediate project requirements by making short-term adjustments to project assignments over a horizon of a few months. *Medium-term resource planning* is based on anticipated project resource needs in addition to current project assignments, and its focus is on balancing resource capacity and demand by changing the number, timing, scope, or mix of projects in the pipeline, or by increasing capacity to accommodate anticipated needs. *Long-term resource planning* is based on resource estimates and has a time horizon of a year or more, depending on the product development lifecycle. These are aggregate estimates made for strategic planning purposes, as opposed to pipeline balancing purposes.

These three time horizons of resource planning are illustrated in Figure 6–1. Short-term resource planning based on assignments is generally very accurate for two to three months, since it describes what everyone is currently assigned to do. It begins to lose accuracy beyond this time horizon because there is no visibility of potential assignments. Medium-term resource planning tends to be accurate for the duration of current projects because it includes

F I G U R E 6–1

Resource Planning Time Horizons

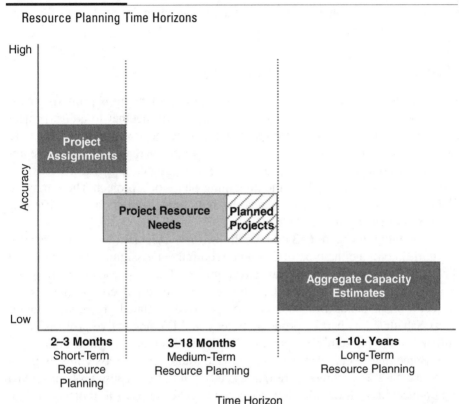

planned resource needs for subsequent project phases of all projects in development. It's necessary to combine project resource needs with project assignments to provide a complete picture of resource planning, since the two always overlap to some degree. In addition, some companies can extend project resource needs planning even further by including the needs of planned projects not yet started. We'll discuss planned products in Section Four.

Long-term resource estimates can extend visibility for years into the future. Typically, the source of the data on which these estimates are based is some form of long-term resource modeling, rather than the derivation of estimated needs from project information.

While we focus on medium-term capacity planning in this chapter, we will also discuss short-term and long-term capacity planning to provide the necessary context. We'll discuss three practices of medium-term capacity planning: capacity planning by resource groups, integration of capacity planning with annual financial planning, and capacity planning that takes place as part of pipeline and portfolio management.

Short-Term Resource Capacity Planning

As we discussed in Chapter 5, short-term capacity planning is primarily tactical. Resource managers and project managers work together to adjust project assignments in order to increase utilization and cover project needs. There are two types of project assignments: those that are planned and those that are approved. In companies applying a phase-review project-approval process, resources are only approved for the upcoming phase of a project. There may be planned assignments for future phases, but these need to be identified as planned, not as approved.

The table in Figure 6–2 defines the various categories for resources. The categories combine the type of resource requirement (assignment or need) and the status of the project phase (approved phase, planned phase of an approved project, or a planned project). Usually, a phase-review decision changes the project's phase status from planned to approved. Planned projects are tentatively scheduled as part of product strategy and become active projects when initiated in the phase-review process. We'll discuss this in Section Four.

Sometimes a company will try to make future resource assignments in order to increase its visibility of resource needs, but this can cause problems. On a long project, it's a little ridiculous to assign R&D developers to projects far into the future, such as assigning Mary Smith and Thomas Latham from the Test Department to staff project work starting two years from now in an effort to get

F I G U R E 6–2

Definition of Resource Categories and Approval Status

	Approved Phase	Planned Phase	Planned Project
Resource Assignments	Approved Assignment	Planned Assignment	NA
Resource Needs	Approved Need	Planned Need	Planned Project Need

visibility of the project's future resource needs. Similarly, it's misleading to assign an unnamed resource such as "To Be Determined," because that confuses demand with capacity, assuming an additional resource that doesn't exist.

In some environments, short-term resource capacity planning is sufficient. If projects are short and all resources are assigned for the entire project at the beginning, then assignment information is complete. Some IT organizations run their projects this way. They accept project requests continually, prioritize the queue of requests, and assign resources as they become available. In these cases, capacity is managed by allocating it as it becomes available. However, in product development, capacity management is more complex. Resources are not all assigned for the duration of the project, and project approval is usually done phase by phase, as decisions are made about continuing, postponing, or canceling projects. In these cases, medium-term capacity planning is critical.

Medium-Term Resource Capacity Planning

Medium-term resource capacity planning adds planned project resource needs to project assignments in order to provide additional visibility over a longer planning horizon. Typically, medium-term resource capacity planning can increase visibility of resource demand to 12 months or more. For many companies, a 3 to 18-month period is the most effective time frame for making decisions about how to balance resource capacity with project demand. Adjustments can be made to delay or cancel projects in a portfolio in order to balance resource demand with capacity, and adjustments also can be made to increase capacity to meet the resource demand through hiring or reassignments.

With complete resource-need estimates for all projects, it is feasible to compare the resource needs of a project portfolio to available capacity. This is typically done at an aggregate level for all R&D resources and for selected resource groups that may be constraining the pipeline.

Figure 6–3 illustrates resource demands (approved assignments, planned assignments, and planned needs) compared to capacity (planned headcount) for the Software Engineering Group at Commercial Robotics. CRI uses a phase-review project-approval process and maintains the distinction between approved assignments for phases approved by its Product Approval Committee and planned assignments for yet-to-be-approved phases of approved projects. This distinction enables everyone to understand the status of resource assignments more clearly and does not undermine the critical authority of the Product Approval Committee or similar executive decision-making group.

Headcount in the Software Engineering Group at CRI is currently 25, with 22 assigned to projects. As Ted Johnson, the Software Engineering Manager,

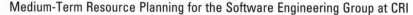

F I G U R E 6–3

Medium-Term Resource Planning for the Software Engineering Group at CRI

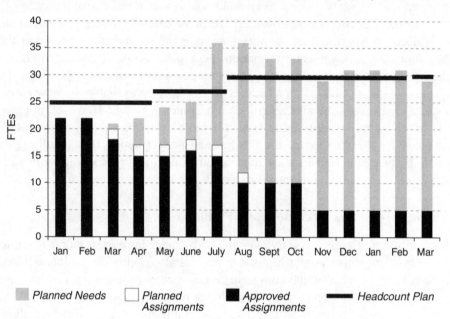

reviews his staff assignments, he is comfortable that he has the resources to fill all project needs through June. He can fill all requested assignments now and still have capacity to meet all planned needs as his developers complete their current project assignments and the new hire starts in May. However, by July he has a big problem approaching. Even with the additional hire expected in August, he does not have the capacity to meet the planned need for 36 software engineers. This problem appears to continue until March of next year, yet Ted knows only the needs that have been projected through the end of the current year. Since his problem probably extends well into the coming year, he prepared a plan to increase the size of his group.

Integration with Annual Financial Planning

Here's a very important point: Medium-term resource capacity planning can be integrated with annual financial planning, *if* resource needs are planned sufficiently in advance to cover the horizon of the annual plan. If, for example, the financial plan is done in the fall for the upcoming year, then all planned projects anticipated to start next year must be added to those already under way. This is represented in Figure 6–1 by the extension of planned projects. If these projects aren't included, then the resource needs calculation will be incomplete and will not be useful in annual financial planning. But when they're added, the result is a significant new planning capability.

Commercial Robotics' CFO, Shaun Smith, was frustrated with how R&D resource plans were previously made as part of the financial plan. "Every year, all R&D departments request a 30 percent increase in their budgets. We need to invest more in R&D, but I was convinced that there must be some resource groups that need to grow faster than others. When I asked for the detailed analysis of expected needs, I got the answer that it can't be estimated. So we really didn't know how to put together our resource budgets."

CRI addressed this problem by extending its medium-term resource planning to cover the same planning horizon as its annual plan. Through its product strategy process, CRI created product platform and product line plans to include all new products expected to be worked on in the upcoming year. These plans then needed to be rationalized to a rough expectation of overall R&D capacity, but, as Shaun said, "This must be done anyway to give us a reasonable forecast for future revenue from new products."

The result of extending medium-term resource planning is that headcount plans for the resource groups can be tied to forecasted needs, and the budget can be based on these assumptions. "For the first time, everyone feels like we have R&D budgets that are based on realistic capacity-planning assumptions," Shaun told the Board of Directors. "Overall, we increased the R&D budget by 15 percent. Some resource groups weren't happy because they got no increase, while others got increases of more than 20 percent, but I believe that this will help us align capacity with our real needs, and that we'll increase overall R&D utilization and output."

Pipeline Capacity Management

Product development pipeline management is a very important element of medium-term resource capacity planning. This is where priorities are set among all projects and potential projects to allocate development capacity in order to achieve strategic objectives. Evaluating project needs compared to capacity for a resource group, as illustrated in Figure 6–3, is an effective way for one group to assess resource demand compared to capacity, but it's more difficult to do this across all groups and all projects.

There are various ways of doing pipeline management; techniques vary according to the sophistication of different processes and systems. Perhaps the

best way to envision pipeline management is as the evaluation of a set of portfolio alternatives against R&D capacity, with the objective of maximizing a
specific financial factor, such as revenue or profit. Each portfolio alternative is
evaluated against capacity to see if it's feasible, and from those feasible alternatives the one that has the highest revenue or profit potential is usually selected. Resource management provides the capacity dimension of this analysis.
We'll discuss pipeline management in detail in Chapter 18, where dynamic
pipeline management is introduced.

Long-Term Resource Capacity Planning

Companies with very long development cycles, such as automotive, pharma-
ceutical, and capital equipment companies, may also find it helpful to plan resource capacity over a long-term horizon of up to 10 years or even longer.
Typically, this capacity planning is carried out in conjunction with long-term
planning over that horizon, since this helps companies to plan the approximate
size of the R&D investment needed to achieve long-term objectives.

Long-term resource capacity planning is best done at a rough-cut aggregate level, determining, for instance, the quantity of high-level skill requirements per year, generally using some sort of modeling approach. The models
vary by type of product, size of company, and approach to planning. In some
industries, long-term resource capacity planning can enable companies to
make the necessary strategic shifts in their product development.

There is generally some value in integrating this long-term resource
modeling with the operational resource management system, and it can be
beneficial to use the same skill categories for both medium-term and long-term
resource planning so that the results can be considered together. It may also be
beneficial in some cases to integrate long-term with medium-term resource demand to create a more complete resource profile.

PROJECT RESOURCE NEEDS

Project resource needs are very different from project assignments, yet this
distinction is sometimes missed, and project assignments are used in place of
resource needs. When this happens, it limits resource flexibility and restricts
the value of medium-term resource planning. It also confuses assignments by
using them for both needs and assignments.

Let's look at the differences between resource needs and project assignments. Project assignments are assignments of specific individuals, but project
resource needs are not. A project resource need is based on a needed skill,

which indicates the type of person required to fill that need. Project assignments are very specific, while project resource needs are general and are refined and changed as the project progresses. For example, the need for eight programmers for Phase 2 of a project is only a planning approximation. It will change when the product is more defined, toward the end of Phase 1.

Project resource needs change in several ways prior to being transformed into assignments. First, a need may be initially defined at a more generic skill level and then refined to more specific skills as the project progresses. For example, the initial project resource need for eight programmers evolves to become one software development supervisor, one senior software developer, three C++ software developers, and three Java software developers. Second, the quantity of a project resource need can change as development requirements become clearer. In this example, the original need for eight developers grew to nine with the addition of another Java developer, but the software development supervisor is only needed half time. So the revised need is 8.5, with a different distribution of skills. Finally, in order to better support the workload and balance the work across other projects, the resource manager decides to delay assignment of two of the four Java developers, since the pair won't be available until two months later, but assigns five at that time to cover the work to be done. The necessary *amount* of resource is thus assigned to the need, but the *number and timing* of the resources have changed.

Project Planning with Resource Needs

Stage 2 capabilities allow project managers to plan future project resource needs more formally by identifying all expected resource needs of their projects as part of an integrated resource planning system. Prior to having an integrated development chain management system, they could only make estimates of future needs in order to budget the expected project cost, but these estimates couldn't be integrated into resource needs for purposes of capacity planning.

The Fast-Food Robot project at CRI illustrates how a project manager typically plans resource needs. Anne Miller is currently in Phase 1 of her project and is planning the resource needs for the project's subsequent phases, as illustrated in Figure 6–4. She continues to assume that she will work full time on the project, so she includes herself as project manager. She does likewise with Richard Salisbury, extending his assignment to half time (0.5) for the duration of the project. Anne realizes that this does not assure that Richard will continue to be assigned to fill the remainder of the project need, but by indicating this assignment of him as a planned assignment attached to the current approved assignment she is indicating her preference.

FIGURE 6-4

Project Resource Assignments and Needs for the Fast-Food Robot Project at CRI

Category	Resource Assigned	Phase 1 Planning Approved							Phase 2 Development Planned									Phase 3 Test Planned		
		Jan	Feb	Mar	Apr	May	June	Total	July	Aug	Sept	Oct	Nov	Dec	Jan	Feb	Mar	Apr	May	June
Project Management																				
	Miller	1.00	1.00	1.00	1.00	1.00	1.00	6.00	1.00	1.00	1.00	1.00	1.00	1.00	1.00	1.00	1.00	1.00	1.00	1.00
Marketing																				
Marketing Manager	Salisbury	1.00	1.00	0.50	0.50	0.50	1.00	4.50	0.50	0.50	0.50	0.50	0.50	0.50	0.50	0.50	0.50	0.50	0.50	0.50
Software																				
Software Design	Takeyama	0.50	0.50	1.00	1.00	1.00	1.00	5.00	1.00	1.00	1.00	1.00	1.00	1.00	1.00	1.00	1.00	1.00	1.00	1.00
Software Developer	Krieger	0.00	0.00	1.00	1.00	1.00	1.00	4.00	1.00	1.00	1.00	1.00	1.00	1.00	1.00	1.00	1.00	1.00	1.00	1.00
Programmer									8.00	8.00	8.00	8.00	8.00	8.00	8.00	8.00	8.00	8.00	8.00	8.00
Finance																				
Finance Manager	Hall	0.50	0.50	0.50	0.50	0.50	0.50	3.00	0.25	0.25	0.25	0.25	0.25	0.25	0.25	0.25	0.25	0.25	0.25	0.25
Financial Analyst	Bailey				1.00	1.00	1.00	3.00									1.00			1.00
Electrical Engineering																				
Electrical Design	Tan	0.25	0.25	1.00	1.00	1.00	1.00	4.50	1.00	1.00	1.00	1.00	1.00	1.00	1.00	1.00	1.00	1.00	1.00	1.00
Electrical Engineer									4.00	4.00	4.00	4.00	4.00	4.00	4.00	4.00	4.00	4.00	4.00	
Mechanical Engineering																				
Mechanical Design	Stevens	0.25	0.25	1.00	1.00	1.00	1.00	4.50	1.00	1.00	1.00	1.00	1.00	1.00	1.00	1.00	1.00	1.00	1.00	1.00
Mechanical Engineer	Welch			1.00	1.00	1.00	1.00	4.00	1.00	1.00	1.00	1.00	1.00	1.00	1.00	1.00	1.00	1.00	1.00	1.00
Mechanical Engineer									2.50	2.50	2.50	2.50	2.50	1.00	1.00	1.00	0.50			
Quality																				
Quality Manager	Ingram	0.25	0.25	0.25	1.00	1.00	1.00	3.75	0.50	0.50	0.50	0.25	0.25	0.25	0.25	0.25	0.50	1.00	1.00	1.00
Quality Analyst																	0.50	2.00	2.00	2.00
Manufacturing																				
Manufacturing Engineer									1.00	1.00	1.00	1.00	1.00	1.00	1.00	1.00	1.00			
Procurement												0.50	0.50	0.50	0.50	0.50	0.50	0.50	0.50	0.50
Test																				
Test Engineer														1.00	1.00	1.00	4.00	4.00	4.00	4.00
Test Supervisor																0.50	0.50	0.50	0.50	0.50
Project Total		3.75	3.75	7.25	9.00	9.00	9.50	42.25	22.75	22.75	22.75	23.00	22.50	22.00	22.25	21.75	28.25	27.25	27.25	24.25
Project Cost (000)		$54	$54	$105	$131	$131	$138	$613	$330	$330	$330	$334	$326	$319	$323	$315	$410	$395	$395	$352

Anne does the same with Naoto Takeyama and David Krieger in Software. She knows that her project will require significant programming work, but does not yet know exactly how many programmers and what skills will be needed. When the design specification is finished, she will be able to define her programming needs much better. Nevertheless, she wants to give the software engineering manager some indication of her project's potential needs. He asked her to plan her needs at the more generic programmer skill level and then be more specific when she was able to. The approach she took was better than guessing and then changing the skill sets required.

Although most of the finance work would be completed in Phase 1, Anne felt that she needed ongoing financial support, so she indicated this need at one-quarter time (0.25 FTE), with a planned assignment that Art Hall would continue to play this role. Alan Tan would continue full time as the electrical designer, but she planned on adding four electrical engineers in Phase 2 and the first two months of Phase 3. Lynda Stevens and Harold Welch worked with her to come up with the need estimates for mechanical engineers in Phase 2. The need would be heavier in the first half of the phase and then decline in the second half. At this point, the resource requirements for the Test Phase are only approximate, so Anne uses the previously established guidelines for projects like hers.

Once she aggregated her expected resource needs for all post-Phase 0 project phases, Anne estimated the resource cost for the project at $5.8 million. This was under the preliminary estimate she used in Phase 0 to justify the project, but still uncomfortably close to her initial estimate and leaving little margin for error. She wondered how to position this estimate at the upcoming Phase 1 approval meeting.

Resource Needs and Skill Categories

Resource needs provide a powerful way to communicate anticipated resource requirements without frustrating everyone by making premature assignments. However, since resource needs are not planned by person, there has to be some other basis for identifying type of need, usually a skill category of some sort.

A skill category, or capacity planning skill, is a useful way to describe the type of person who will be needed to do the work on a project. Skill categories are critical to effective medium-term resource planning based on project needs. If they are defined incorrectly, the wrong capacity may be planned or evaluated. If the categories are too detailed, then it will be impossible to match anyone to them. If they are too general, they become useless as a planning tool.

There is also confusion between skills and skill categories used for resource planning. A person may have multiple skills, but it becomes difficult to do capacity planning if the plan calls for each person to use those skills. For example, if a software group has five software programmers and they each have a skill for JAVA and C++, how do you define the group's capacity? Is it five resources at each skill for a total of 10 FTEs, even though there are only five people? Do you arbitrarily say there are 2.5 FTEs of each skill and limit flexibility?

The solution is to design skill categories around the most important capacity planning skills. These categories can be defined at an aggregate skill level, or for specific constraining skills. This is one of the most critical factors in implementing medium-term capacity management.

Resource Group Management with Resource Needs

Stage 2 capabilities also give resource group managers extended visibility of planned resource needs beyond approved and planned assignments. In many cases, the effective visibility is extended from a couple of months to as much as a year. With Stage 1 capabilities, resource group managers only had visibility of assignments, and had to make decisions based only on this limited horizon. With Stage 2 capabilities, they can see much further ahead, and can use this additional visibility to make better scheduling and assignment decisions.

Figure 6–5 illustrates Ted Johnson's visibility of resource demand and capacity for his group in terms of planned needs (top half) and available capacity net of assignments (bottom half). He can see the planned resource needs for his group in four categories: software design, software development, programming, and JAVA programming, a subcategory of programming. He now has visibility of the eight planned needs for programmers in July for the Fast-Food Robot project that Anne just added. Ted can also see that the ID Robot project just requested resource assignments for one software developer and three programmers.

In the bottom half, Ted can see all the current assignments of his people. These assignments define his available capacity. So he decides to expand the assignment detail for David Krieger, scheduling his approved assignment to the Fast-Food Robot project as well as the *planned assignment* starting in July. The system also shows Ted where the overassignments are and what the total availability is. (The chart only includes 5 of Ted's 25 people, but the totals include all 25.)

Ted begins to work on filling the needs that were requested. The software developer request from the ID Robot project is ideal for Susan Collins, and she will be available in March, so he assigns her to this need. The ID Robot project also requested three programmers, and he thinks he can easily fill

Software Engineering Resource Group Manager's View of Resource Demand (Needs) and Capacity (Availability Net of Assignments) at CRI

Skill	Needs/Requests	Jan	Feb	Mar	Apr	May	June	July	Aug	Sept	Oct	Nov	Dec
Software Design	Industrial Ultra				1.00	1.00	1.00	1.00	1.00	1.00	1.00	1.00	1.00
	Adv. Monitoring System					1.00	1.00	1.00	1.00	1.00	1.00		
Software Development	ID Robot (Requested)			1.00	1.00	1.00	1.00	1.00	1.00	1.00	1.00	1.00	1.00
	Special #145					1.00	1.00	1.00	1.00				
	Material Tester V3							4.00	4.00	4.00	4.00	4.00	4.00
Programmer	Fast-Food Robot							8.00	8.00	8.00	8.00	8.00	8.00
	ID Robot (Requested)				3.00	3.00	3.00	3.00	3.00	3.00	3.00	3.00	3.00
JAVA Programmer	Hotel Cleaner								3.00	3.00	3.00	3.00	3.00
	Industrial Robot V2								2.00	2.00	2.00	4.00	6.00
Total		0.0	0.0	1.0	5.0	7.0	7.0	19.0	24.0	23.0	23.0	24.0	26.0

Name	Assignments	Jan	Feb	Mar	Apr	May	June	July	Aug	Sept	Oct	Nov	Dec
Naoto Takeyama	Available (Overassigned)	0.00	-0.75	0.00	0.00	0.00	0.00	1.00	1.00	1.00	1.00	1.00	1.00
David Krieger	**Projects:**												
	Fast-Food Robot			1.00	1.00	1.00	1.00	1.00	1.00	1.00	1.00	1.00	1.00
	Industrial Software V10	0.50	0.50										
	Nonproject:												
	Training							1.00					
	Server Upgrade	0.50											
	Available (Overassigned)	0.00	0.50	0.00	0.00	0.00	0.00	-1.00	0.00	0.00	0.00	0.00	0.00
Jennifer Jones	Available (Overassigned)	0.00	0.00	-1.00	0.00	0.00	0.00	-1.00	-0.50	0.00	0.00	0.00	0.00
Harry Oh	Available (Overassigned)	-0.50	-0.50	-0.50	0.00	0.00	0.00	-1.00	0.00	0.00	0.00	0.00	0.00
Susan Collins	Available (Overassigned)	0.00	0.50	1.00	1.00	1.00	1.00	0.00	0.50	1.00	1.00	1.00	1.00
Total	Scheduled on Projects	22.0	20.0	18.0	15.0	15.0	16.0	15.0	10.0	10.0	10.0	5.0	5.0
	Available	0.5	2	5	8	8	6.5	5	5	13	11.5	15.5	15.5

them in April, since he has eight people available. But now that he can see that the Fast-Food Robot needs eight programmers in July, he anticipates a problem. He knows that the Fast-Food Robot is a top priority, and that he won't have sufficient resources to support it if he gives all eight of his available programmers to the ID Robot project. He decides to formally raise this issue with the Product Approval Committee, which also has visibility of the impending programmer shortage. "Previously, I would not have had visibility of upcoming project needs," Ted explains, "and I would have assigned resources on a first-come-first-served basis, and then we would have re-assigned staff from project to project. Now we can make much better decisions based on visibility over a longer horizon."

Ted also anticipates future resource problems where demand will exceed capacity. In this example, as shown in Figure 6–5, demand would exceed capacity in July (19 FTEs needed, compared to 5 available). So Ted made this problem known to portfolio managers and senior executives who might need to cancel or defer projects because of this constraint. He also began to explore potential alternatives to mitigate this problem, such as identifying contractors who could supplement the capacity of his group. In particular, he worked closely with Anne Miller to come up with a solution prior to her team's Phase 1 review.

RESOURCE TRANSACTION PROCESS

The process of requesting and assigning resources to projects is referred to as the resource transaction process, and the total number of resource transactions can be quite large. For example, in a company like CRI, with 600 developers who are assigned to multiple projects and phases within a year, there could be several thousand resource assignments. With each project need requiring a request and an assignment, there could be 5,000 or more resource transactions in a given year.

Stage 2 capabilities allow a company to formalize its process for filling resource requests with actual assignments. This is sometimes classified as a Stage 2B capability because some companies may prefer to implement the previously described aspects of Stage 2 prior to implementing a resource transaction process.

A resource transaction process establishes the formal organizational authority and responsibility for resource management that was not supported by Stage 1 or Stage 2A systems or processes. With more advanced resource management processes, a project manager will no longer have the authority to directly make resource assignments. She can plan resource needs and make a request from resource managers to fill those needs. In some cases, a specific resource may even be requested, but it is up to the resource manager to make the final assignment.

With a Stage 1 level of capability, these resource transactions could only be managed informally, with the resulting problems of inefficiency, inconsistency, and frustration. Project managers frequently needed to chase resource managers to get assignments filled. The informality was also very inefficient for resource managers. They couldn't see the profile of expected needs and make intelligent assignment decisions for a group of needs. One research study by PRTM showed that it took resource managers from one week to three months to fill a project assignment request, and that they spent 10 percent of their time administering these requests.[1]

A formal resource transaction process increases the efficiency of requesting and making project resource assignments. In its simplest form, the process can be viewed as the series of transactions described in Figure 6–6.

A formal resource transaction process requires a system for managing the transactions as well as a management process defining the "rules of engagement" for requesting and making assignments. The transaction management system must be integrated with the broader development chain management system for greatest effectiveness. It's helpful to understand that a resource transaction process consists of three types of transactions—needs, requests, and assignments—for each serves a different purpose.

Resource Needs

In the resource transaction process, resource needs provide initial visibility of a future assignment request. The need describes the expected skill category required, which may become more specific as the time comes for the need to be changed to a request. This enables the resource manager to better plan resource allocation to fit collective planned needs. A resource need generally does not allocate capacity, but simply provides visibility of a resource requirement. Action is taken on a resource need when that need becomes a request.

Resource Requests

Typically, the filling of a resource need is triggered by a request from the project manager. For example, a project manager working on resource planning for the upcoming phase of a project will review and revise resource needs for that phase, and then issue requests to resource group managers to fill these needs with assignments. With advanced DCM systems, requests can be automatically

F I G U R E 6-6

Resource Transaction Process

Resource Transaction Process
Project managers and resource managers work together to manage the initiation of resource requests from project resource needs, the response to these requests with assignments, and the coordination of all exceptions, changes, and adjustments to these resource transactions.

Project Data
Project managers enter data on anticipated project resource needs, the skills of those needs, any special requirements, and in some cases individuals requested.

Resource Data
Resource managers maintain availability data on their people and identify their skills.

  Project Plan

  Project Budget

 Project Resource Plan

XXX XX XX
XXXX XXXX
XXXXXXX Project Work Requirements

 Resource Availability   Projected Resource Needs

 Utilization Measurement   Resource Group Skills

Resource Management System
Maintains project resource assignments and needs, and consolidates these across all projects and resource groups.

Resource Assignment
Initiates assignment request from resource need and matches need to available resources with specified skill.

Project Manager
Determines resource need to support project based on availability of that resource skill, and submits request for assignment.

Resource Manager
Receives request for assignment, reviews resource availability, and makes project assignment.

issued for all the resource needs in a phase. A project may also entail continuing planned assignments that go into a yet-to-be-approved phase, and these will be re-affirmed by the resource manager as part of this request process.

In companies that follow a phase-based approval process, resource requests and the assignments to fill them do not yet mean that the resources are approved. The resources are assigned pending approval of the phase, and if, for some reason, the project phase isn't approved, then the resource assignments are canceled and the resources are available for other projects.

Resource Assignments

Resource managers respond to requests in a variety of ways, and resource requests may be filled differently than originally requested, due to constraints on availability. For example, a resource manager may be able only to partially fill a resource request; e.g., perhaps there is only availability of one half-time person in March despite the request for one full-time person. It's possible that a request may be filled for fewer resources than requested, or may be filled with a different mix of skills. In the case of any major discrepancies between what is requested and what is delivered, the project manager and the resource manager will typically discuss the matter and agree jointly what to do, whereas more routine differences may be handled directly through the resource transaction system. The boundaries for exceptions are defined by the formal resource-transaction management process.

Project resource assignments can also be changed in the course of a project if a developer leaves the company or is reassigned to a higher-priority project as part of pipeline and portfolio adjustments. The resource transaction process therefore needs to handle mid-project resource adjustments such as these.

REQUIREMENTS

Requirements for implementing Stage 2 capabilities generally build upon those of Stage 1. Here again, similar to resource assignments, it's necessary to include the complete resource needs of all projects, as well as the capacity of all resource groups within a business unit.

To achieve the capabilities outlined, companies need a resource management system that automates project resource needs in addition to resource assignments. Otherwise, they'll be forced to use some restrictive management practices, such as predetermining resource assignments long before it's appropriate. The resource management system should be capable of filling project needs to create assignments at appropriate times.

Stage 2 performance capabilities, such as medium-term capacity planning, are supported by the broader capabilities of a DCM system. This includes enabling resource managers to compare resource demand (needs) to capacity, to evaluate gap analysis for selected skills, and to provide resource managers with the tools to manage medium-term capacity. Under advanced DCM systems, pipeline capacity management can be integrated with portfolio and product strategy. Long-term capacity management may or may not be implemented at the same time as other Stage 2 capabilities. There may also be a value in integrating long-term capacity management with the primary resource management application.

Companies that follow a phase-review process like PACE® authorize resources for projects phase by phase. Those that follow other phase-review or stage-gate processes that don't provide for phase-by-phase authorization can now add this capability. They should make sure, however, that the DCM system supports this practice, or else it will be undermined. If the system doesn't recognize the approved status of resource assignments, an organization can expect that resources will be assigned for the duration of a project at the beginning, because this is how the system reports it.

At Stage 2B, a company automates the resource transaction process, and the DCM system should integrate this functionality with medium-term capacity management. This system would include the ability to handle needs and requests from project managers and assignments against these needs by resource managers.

There are two major management process requirements at the Stage 2 level of capability. The first is the requirement to establish skill categories, because project resource needs are based on skills. Definitions of these skills and categories must be devised carefully; otherwise, resource planning will be less effective. Skill categories need to be mapped to specific resources, since this is how a need is transformed into a specific assignment.

The second major management process requirement is a formal resource transaction-management process. This process defines how resource needs are identified, how assignments are requested based on these needs, and how assignments are made from these requests. It's important that these transactions are clearly defined, and that all project and resource managers are appropriately trained in order to avoid internal resistance to the new processes and applications. Companies that have implemented a resource transaction system without defining and implementing the associated management process are generally frustrated by the difficulties that arise.

Several organizational responsibilities need to be clarified in Stage 2. The responsibilities and authorities of project managers and resource managers within the resource transaction process are critical. These cover a broad range of situations, such as attrition during a project assignment, special resource assignment requests, communication of needs, accuracy of resource-availability data, guidelines for skill definitions, etc.

In addition, it's important to clarify responsibilities and expectations for medium-term capacity management. What is expected of resource group managers? Who is responsible, and what is the process for resolving future capacity conflicts? How is capacity planning integrated with annual resource group budgeting?

BENEFITS

Stage 2 builds on the initial benefits of Stage 1 by extending the capacity planning horizon from the short term to the medium term, further increasing the effectiveness of resource management. Stage 2 also automates the resource transaction process, providing much more efficient administration of resource requests and assignments. The benefits of Stage 2 resource management include the following:

1. Product development output is increased because resource capacity is better fit to resource needs through medium-term resource capacity management and pipeline management.

2. Utilization is further improved through better medium-term resource capacity management.

3. Increased visibility of resource availability for the medium term enables project managers to extend resource-based project planning for the duration of their projects, which increases the effectiveness of project planning.

4. With increased visibility of consolidated resource needs, fewer projects are likely to be started only to be canceled because of future resource constraints.

5. The automated resource transaction process reduces resource administration for project managers and resource managers, freeing them for more important value-added work.

6. A formal resource-transaction process reduces conflict among project managers and resource managers regarding resource needs and expectations.

7. Extending medium-term resource planning through the annual planning horizon enables a company to integrate resource capacity planning with the annual financial plan.

SUMMARY

Companies with Stage 2 process and systems capability extend resource management from a short-term to a medium-term horizon by introducing resource needs for future phases of a project. We reviewed the characteristics of the resource planning horizon, and suggested how resource planning integrates with annual financial planning. We explained medium-term resource planning by project managers and how resource group managers use this planning to do resource capacity management. This was illustrated with an example from our hypothetical company, CRI Inc. Finally, we discussed resource transaction management, sometimes referred to as Stage 2B.

Every company whose competitiveness depends on a continuous output of innovative new products contains what might be described as a continuously operating internal market in development resources—chiefly its R&D personnel. We saw earlier how companies at Stage 0 capability (informal resource management) essentially operate a freewheeling internal auction house in which development projects vie with one another for R&D resources in what amounts to open bidding. Stage 1 capability (short-term utilization management) confers the benefits of rudimentary market discipline, as resources are formally assigned and R&D utilization emerges as a distinct organizational goal. The internal market in development resources is no longer left to regulate itself. At Stage 2 (medium-term capacity planning and management), the allocation of development resources to development projects becomes distinctly more farsighted, more systematic, and better regulated.

7

CHAPTER

Stage 3—Project Resource Requirements Planning and Management

At Stage 3, the focus shifts to the *individual project*, with the objective of optimizing the efficient use of resources to successfully complete individual projects. In this sense, the focus of Stage 3 is more specific than that of either preceding stage, for in Stages 1 and 2 the focus is on resource management *across all* projects. Assignments and needs are collected at a high level (Level 1) from all projects and used to manage all resources. The objective is optimization of resources across all projects (short-term optimization in Stage 1 and medium-term in Stage 2), with little concern for how much resource any project consumes.

With a Stage 3 level of capability, companies gain significant benefits: If a project can be completed with 20 percent fewer resources, then the return on investment will increase dramatically. Although we used this illustration earlier, it's worth repeating here. Take a project with a $6 million development cost, including a $5 million cost for developers, and an expected $9 million profit contribution (50 percent ROI above the development cost). Through improved resource requirements planning and management, the resource cost can be reduced to $4 million, and the project cost can be reduced to $5 million. As a result, the project ROI improves to 80 percent [$9M – $5M / $5M].

For some projects, reduced resource costs through resource requirements planning can be the difference between being below or above the hurdle rate required for project approval. With lower project costs, a company can develop more projects. Stage 3 capabilities confer project portfolio benefits as well. A 20 percent lower resource need by all projects means that 20 percent more projects can be completed with the same total resources—with a resultant overall output increase of 20 percent as well.

In this chapter, we'll explore the DCM systems and project resource planning techniques that can help project managers achieve these benefits.

Stage 3 has four major practices, illustrated in Figure 7–1. *Project resource requirements planning* (RRP), which is the planning aspect of Level 2, enables project managers to effectively plan resource requirements. Generally, this is done by providing each project manager with a tool set to make preliminary resource estimates in person-days for each project step. These estimates could come from the project manager's personal experience, they could be based on standards, or they could be derived from a detailed workplan for each project step. The project manager can then balance these resource estimates by comparing them to resource availability, by time-phasing steps differently, or by balancing the estimates among skills.

The second practice associated with Stage 3 is the ability to *translate resource requirements* into needs for purposes of establishing resource plans and requesting assignments. This is the operational end of resource management. Tools like the ones described earlier help the project manager translate resource estimates in person-days into FTEs for resource needs. A project manager may adjust these estimates based on the number of workdays in the months being planned, and may round off the needs. For example, the need for 22 person-days in a 20-day month computes to 1.1 FTEs, but it's more practical to request a single person (1.0). Finally, the project manager may find it useful to adjust her needs estimates to fit the availability of specific individuals being requested for their skill sets. To do this, she must have the information created at the earlier stages of capability we've described.

The third practice is *project resource requirements management* (RRM), shown on the right side of the framework (Figure 7–1). RRM enables project managers to effectively manage resource assignments by allocating assignments to project steps. This allocation may be similar to the original resource-requirement estimates, but could be different, depending on how needs were filled and the specific developers assigned to a project. Typically, a project manager finds it helpful to make adjustments in order to make the best use of the abilities and interests of those actually assigned to projects. At this point, the project manager also has the opportunity to create a resource contin-

Overview of Stage 3 Practices

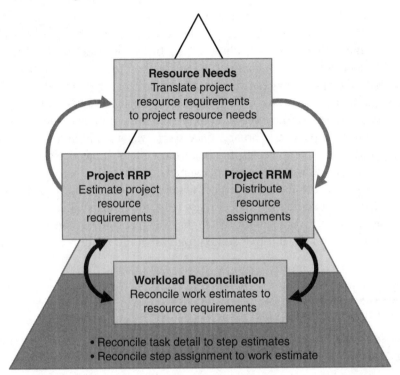

gency by allocating some of the assigned resource time to a buffer, instead of fully allocating it to all steps. This enables the project manager to implement some of the resource aspects of critical chain planning.

The fourth and final practice associated with Stage 3 development resource management is *workload reconciliation*—the reconciliation between Levels 2 and 3. This practice enables project team members to reconcile detailed work scheduling estimates to their step assignments. It also enables the reconciliation of work estimates from detailed step workplans to step resource requirements. Without such tools and visibility, managers cannot possibly manage time frames, tasks, steps, etc.

To better understand each of these components of project resource requirements planning and management, we'll look at how Anne Miller applied them when she did the Phase 1 planning for the Fast-Food Robot project at our hypothetical company, CRI.

PROJECT RESOURCE REQUIREMENTS PLANNING (RRP)

In preparation for the Phase 0 review, Anne started planning the major steps for Phase 1, using the standard guidelines for a project of this type. Phase 1 consists of the six steps illustrated in Figure 7–2. Project management is a step that continues throughout the entirety of Phase 1. The first major step is a market study, which included the requirements definition for the product. Anne thought this step would take approximately three months. Marketing had already done a lot of strategic work, but her project team needed to translate that work into specific product characteristics. She expected that those on the project could start developing the business plan in February, once work was already under way for the market study. She also expected to spend most of the first phase refining the business plan, which was the major deliverable of Phase 1.

Anne planned to have her project team start work on the Fast-Food Robot's functional specification in February. She thought that the work on that specification could overlap with the high-level mechanical and software design steps of this phase, but that some progress would need to be made on the

F I G U R E 7–2

Fast-Food Robot Phase 1 Steps

functional specification before work could begin on the mechanical and software design steps. Overall, Anne expected that six months would be required to complete the work of Phase 1—the standard for a project of this complexity at CRI. So Phase 1 would be completed by the end of June.

Using Resource Requirement Guidelines

In some cases, the project manager may use recommended guidelines for estimating resource requirements. For example, a company may have established resource guidelines for standard steps, which estimate the amount of resources typically needed, by skill, for a step. In some cases, these guidelines may be referred to as resource standards because they are used in a similar way to direct labor or cost standards. These guidelines can come from either manual standards or automated step libraries that plug the resource estimates directly into the step for planning purposes.

As Anne planned Phase 1 of the Fast-Food Robot project, she actually selected standard project steps from an automated step library. Resource profile estimates that specified the resource requirements guidelines for each skill category were attached to each of these steps. This is illustrated in Figure 7–3. The project management step, for example, had a guideline for the five members of the core team at five days per month for each core team member and 10 days per month for the project team leader.

Anne used the resource estimates from the step library as a starting point, even though she knew this project had some special characteristics and expected that she would need to make adjustments to them based on her own experience. Nevertheless, the standard estimates from the library provided her with good input, and she didn't need to start with a blank piece of paper. The automated tool from CRI's DCM system that Anne was using to do her project management had this step library embedded, so it was easy to use.

Making Preliminary Resource Estimates

Anne then used the resource estimating tool to prepare her preliminary resource estimates from these guidelines. Figure 7–4 illustrates how she made these estimates. The worksheet is organized by step within Phase 1, and includes estimates for each skill category for each step. Estimates are stated in the number of person-days per month.

First, Anne revised the estimate for the marketing manager's time in January from 20 person-days to 10 person-days. The original estimate of 60 days was spread evenly over three months, but, with the adjustment, the total was

F I G U R E 7–3

Steps and Resource Profile for Phase 1 of a Project

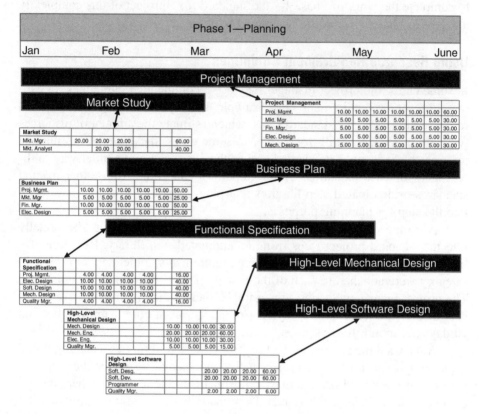

Phase 1—Planning					
Jan	Feb	Mar	Apr	May	June

Project Management

Market Study

Market Study
Mkt. Mgr.	20.00	20.00	20.00				60.00
Mkt. Analyst		20.00	20.00				40.00

Project Management
Proj. Mgmt.	10.00	10.00	10.00	10.00	10.00	10.00	60.00
Mkt. Mgr	5.00	5.00	5.00	5.00	5.00	5.00	30.00
Fin. Mgr.	5.00	5.00	5.00	5.00	5.00	5.00	30.00
Elec. Design	5.00	5.00	5.00	5.00	5.00	5.00	30.00
Mech. Design	5.00	5.00	5.00	5.00	5.00	5.00	30.00

Business Plan

Business Plan
Proj. Mgmt.	10.00	10.00	10.00	10.00	10.00	50.00
Mkt. Mgr	5.00	5.00	5.00	5.00	5.00	25.00
Fin. Mgr.	10.00	10.00	10.00	10.00	10.00	50.00
Elec. Design	5.00	5.00	5.00	5.00	5.00	25.00

Functional Specification

High-Level Mechanical Design

Functional Specification
Proj. Mgmt.	4.00	4.00	4.00	4.00	16.00
Elec. Design	10.00	10.00	10.00	10.00	40.00
Soft. Design	10.00	10.00	10.00	10.00	40.00
Mech. Design	10.00	10.00	10.00	10.00	40.00
Quality Mgr.	4.00	4.00	4.00	4.00	16.00

High-Level Software Design

High-Level Mechanical Design
Mech. Design	10.00	10.00	10.00	30.00
Mech. Eng.	20.00	20.00	20.00	60.00
Elec. Eng.	10.00	10.00	10.00	30.00
Quality Mgr.	5.00	5.00	5.00	15.00

High-Level Software Design
Soft. Desg.	20.00	20.00	20.00	60.00
Soft. Dev.	20.00	20.00	20.00	60.00
Programmer				
Quality Mgr.	2.00	2.00	2.00	6.00

now 50 person-days. Anne found that, on a first pass, the guidelines seemed reasonable.

She also developed a detailed task-level (Level 3 in Figure 7–1) workplan for the business plan step in order to determine, from the bottom up, the workload required. Doing this can be useful in selected cases where there is little experience with a particular step and when the estimates need to be derived by "engineering" the step instead of using experience. However, doing this is not recommended for all steps, for it requires a lot of work that can sometimes be of little value, since detailed bottom-up work estimates are frequently overestimates, and frequently incomplete.

In Figure 7–4 the two columns at the right indicate the resource estimate guidelines from a standard (Std.) and from a detailed (WP) workplan. The

F I G U R E 7–4

Fast-Food Robot Preliminary Resource Estimate

Fast-Food Robot										
	Phase 1 Step Planning Person Days								Reference Estimates	
Step Skill Category	Jan	Feb	Mar	Apr	May	June	Total	Est.	Std.	WP
Project Management										
Project Management	10.00	10.00	10.00	10.00	10.00	10.00	60.00	60.00	50.00	74.25
Marketing Manager	5.00	5.00	5.00	5.00	5.00	5.00	30.00	30.00	30.00	28.50
Financial Manager	5.00	5.00	5.00	5.00	5.00	5.00	30.00	30.00	30.00	24.50
Electrical Design	5.00	5.00	5.00	5.00	5.00	5.00	30.00	30.00	30.00	73.50
Mechanical Design	5.00	5.00	5.00	5.00	5.00	5.00	30.00	30.00	30.00	28.50
Market Study										
Marketing Manager	10.00	20.00	20.00				50.00	60.00	80.00	
Marketing Analyst		10.00	20.00				30.00	40.00		
Business Plan										
Project Management		10.00	10.00	10.00	10.00	10.00	50.00	50.00	10.00	
Marketing Manager		5.00	5.00	5.00	5.00	5.00	25.00	25.00	40.00	18.50
Financial Manager		10.00	10.00	10.00	10.00	10.00	50.00	50.00	50.00	45.50
Electrical Design		5.00	5.00	5.00	5.00	5.00	25.00	25.00	30.00	55.00
Functional Specification										
Project Management		4.00	4.00	4.00	4.00		16.00	16.00	8.00	
Electrical Design		10.00	10.00	10.00	10.00		40.00	40.00	40.00	
Software Design		10.00	10.00	10.00	10.00		40.00	40.00	40.00	
Mechanical Design		10.00	10.00	10.00	10.00		40.00	40.00	40.00	
Quality Manager		4.00	4.00	4.00	4.00		16.00	16.00	20.00	
High-Level Software Design										
Software Design				10.00	20.00	20.00	50.00	60.00	60.00	
Software Developer				20.00	20.00	20.00	60.00	60.00	60.00	
Programmer										
Quality Manager				2.00	2.00	2.00	6.00	6.00	10.00	
High-Level Mechanical Design										
Mechanical Design				10.00	10.00	10.00	30.00	30.00	30.00	
Mechanical Engineer				20.00	20.00	20.00	60.00	60.00	60.00	
Electrical Engineer				10.00	10.00	10.00	30.00	30.00	30.00	
Quality Manager				5.00	5.00	5.00	15.00	15.00	15.00	
Total	40.00	128.00	138.00	175.00	185.00	147.00	813.00	843.00	793.00	

business plan step has a standard of 10 person-days for the project manager (Proj. Mgmt.) and 40 person-days for the marketing manager (Mkt. Mgr.). Anne has a marketing background and intends to play an active role in creating the business plan, so she increased her time to 50 person-days and decreased the marketing manager's to 25.

The workplan for the marketing manager was proposed based on a lower level of involvement, and resulted in an estimate of 18.5 person-days. Anne used 25 person-days as a placeholder for now, but will balance this later. The electrical design engineer wanted to do a detailed workplan and estimated 55 person-days for the business plan step. Anne thought that this was too generous, and decided to use an estimate of 25 person-days instead, which is closer to the standard of 30.

At this point she was comfortable with her preliminary estimates and turned her attention to time-phasing the project steps in order to balance resources with step-specific resource needs.

Time-Phasing Steps to Balance Resources

In this example, Anne is trying to balance resource requirements to a need of only one marketing manager. She knows that the marketing manager resource is in demand, and that only one will be assigned to her project. If more are required, then the project might be delayed.

Figure 7–5 illustrates the impact of extending the market study step by a month. The table on the top shows the preliminary resource estimates for the marketing manager in Phase 1 broken out by step, and the table on the bottom illustrates the change.

The marketing manager requirements are currently estimated at 25 person-days in January and 30 in February, which is obviously not feasible for one person. By extending the Market Study step by a month, Anne is able to spread out the resource requirements for the marketing manager on this step over four months (15, 12, 12, and 11 person-days), instead of over three months. This change enables the requirement to be better balanced to one marketing manager, with 20 person-days estimated in January, 22 in February and March, and 21 in April. Anne does not believe that this schedule change will have a major impact on the project, but she realizes she will need to coordinate the overlap more closely with the business plan.

This same technique can be used to fit a project plan to available resources. For example, if eight software developers are required, but only six are available, then the project manager can extend or delay the resource-constrained project steps to fit the resource availability.

F I G U R E   7–5

Example of Extending a Step to Balance Resources

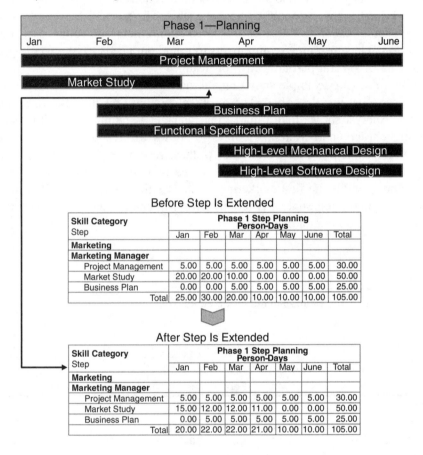

Before Step Is Extended

Skill Category	Phase 1 Step Planning Person-Days						
Step	Jan	Feb	Mar	Apr	May	June	Total
Marketing							
Marketing Manager							
Project Management	5.00	5.00	5.00	5.00	5.00	5.00	30.00
Market Study	20.00	20.00	10.00	0.00	0.00	0.00	50.00
Business Plan	0.00	0.00	5.00	5.00	5.00	5.00	25.00
Total	25.00	30.00	20.00	10.00	10.00	10.00	105.00

After Step Is Extended

Skill Category	Phase 1 Step Planning Person-Days						
Step	Jan	Feb	Mar	Apr	May	June	Total
Marketing							
Marketing Manager							
Project Management	5.00	5.00	5.00	5.00	5.00	5.00	30.00
Market Study	15.00	12.00	12.00	11.00	0.00	0.00	50.00
Business Plan	0.00	5.00	5.00	5.00	5.00	5.00	25.00
Total	20.00	22.00	22.00	21.00	10.00	10.00	105.00

The time-phasing of steps to fit resources will not alter the overall project schedule if the steps are off the critical path, or if the project team can adjust its work, as was the case with the changes Anne made. Time-phasing simply adjusts individual steps to start later than the earliest possible start date, or to continue beyond the earliest possible end date. When a resource-constrained step is on the critical path, then a decision must be made about whether to delay the start of the project, extend the project schedule, find the necessary resources by hiring a person with the appropriate skill set, cancel the project because there are not sufficient resources to complete it, or cancel another project to obtain its resources.

Balancing Resource Requirements among Skills

A project manager can also balance resource requirements to some extent among skills that are reasonably interchangeable. The purpose for doing this is to fit available resources to project resource requirements. In some cases, this is done to balance resource requirements to round out needs to fit a whole person instead of several partial needs, for instance.

The following example shows an alternative way for Anne to resolve the resource conflict just described. The resource summary for the Mkt. Mgr. and the Mkt. Analyst shows that a marketing manager is required for 30 person-days in February and March. Figure 7–6 shows the resource requirements summarized by step in Phase 1 for each of these resources. This would require that a second marketing manager be assigned half time for two months and then dropped, while the first marketing manager's time is reduced to half time in April. A marketing analyst is required for 10 person-days in February and 20 in March to do the market study.

F I G U R E 7–6

Summary Resource Requirements for Selected Skills

Skill Category	Phase 1 Person-Days						
Step	Jan	Feb	Mar	Apr	May	June	Total
Marketing							
Marketing Manager							
Project Management	5.00	5.00	5.00	5.00	5.00	5.00	30.00
Market Study	10.00	20.00	20.00	0.00	0.00	0.00	50.00
Business Plan	0.00	5.00	5.00	5.00	5.00	5.00	25.00
Total	15.00	30.00	30.00	10.00	10.00	10.00	105.00
Marketing Analyst							
Project Management							
Market Study	0.00	10.00	20.00	0.00	0.00	0.00	30.00
Business Plan							
Total	0.00	10.00	20.00	0.00	0.00	0.00	30.00

After reviewing this "picture" of her needs and available resources, and after talking with the marketing manager, Anne decides that some of the market study requirements can be shifted from the marketing manager to the marketing analyst. This shift is accomplished by increasing the marketing analyst requirement to 20 person-days in January, 20 in February, and 20 in March. The requirement for the marketing manager on the market study is reduced to 10 person-days in February and 10 in March. The marketing manager resource requirement for the project phase is now more balanced in February and March. Overall, this is a net increase in requirements by 10 person-days, since the marketing analyst is not as experienced as the marketing manager, but the resource requirements now better fit expected resource availability. Anne is also pleased to see that the cost is lower as well, because the marketing analyst is less expensive than the marketing manager.

TRANSFORMING RESOURCE REQUIREMENTS INTO RESOURCE NEEDS

Once the project manager has estimated resource requirements, she must transform the planned requirements into resource needs to be used for resource scheduling. Remember that resource requirements are stated in person-days, and resource needs are stated in FTEs. Also, resource needs must be filled by real assignments, so it's helpful if requirements are rounded to create needs. It is probably not feasible to completely fill a need that fluctuates month to month, so the project manager has to round-off needs from resource requirements.

Anne Miller, the Fast-Food Robot project manager, is ready to create project needs from the project's resource requirements. How she does this is shown in Figure 7–7. She starts by transforming the requirement for her own time to a need. As you can see, she rounds the requirements of 0.5, 1.2, 1.2, 1.2, 1.2, and 1.0 to an even need of 1.0—in other words, herself at full time.

The 1.2 project manager requirement in February was automatically translated from 24 person-days to 1.2 FTEs [24 / 20]. The total need requirement for the phase was 6.3, so the rounded need is 6.0. This is obviously workable. Anne just needs to balance her work a little.

Because Anne wants Richard Salisbury to be the marketing manager on the project, she adjusts the need to fit his availability: full time for the first two months, half time for the next three months, and full time for the last month. Richard is comfortable that this is workable, even though it's 0.75, approximately 15 person-days [0.75 × 20] less than the estimated requirement.

The software design (Soft. Desg.) skill requirement is estimated at 0.5 FTE in February and March, 1.0 in April, 1.5 in May, and 1.0 in June. Anne realizes that an assignment with this much fluctuation would be inefficient and

Translation of Resource Requirements to Resource Needs

Fast-Food Robot

Department Skill Category		Jan	Feb	Mar	Apr	May	June	Total
				Phase 1 Planning **Resource Needs**				
Project Management	Need	1.00	1.00	1.00	1.00	1.00	1.00	6.00
(Anne Miller)	Requirement	0.50	1.20	1.20	1.20	1.20	1.00	6.30
	Difference	0.50	-0.20	-0.20	-0.20	-0.20	0.00	-0.30
Marketing Manager	Need	1.00	1.00	0.50	0.50	0.50	1.00	4.50
(Richard Salisbury)	Requirement	1.25	1.50	1.00	0.50	0.50	0.50	5.25
	Difference	-0.25	-0.50	-0.50	0.00	0.00	0.50	-0.75
Marketing Analyst	Need	0.00	1.00	1.00				2.00
	Requirement	0.00	1.00	1.00	0.00	0.00	0.00	2.00
	Difference	0.00	0.00	0.00	0.00	0.00	0.00	0.00
Software								
Software Design	Need	0.00	0.50	1.00	1.00	1.00	1.00	4.50
	Requirement	0.00	0.50	0.50	1.00	1.50	1.00	4.50
	Difference	0.00	0.00	0.50	0.00	-0.50	0.00	0.00
Software Development	Need	0.00	0.00	0.00	1.00	1.00	1.00	3.00
	Requirement	0.00	0.00	0.00	1.00	1.00	1.00	3.00
	Difference	0.00	0.00	0.00	0.00	0.00	0.00	0.00
Total	Need	2.00	3.50	3.50	3.50	3.50	4.00	20.00
	Requirement	1.75	4.20	3.70	3.70	4.20	3.50	21.05
	Difference	0.25	-0.70	-0.20	-0.20	-0.70	0.50	-1.05

most likely rejected by the software group. So she adjusts and smooths the need to half time in February and full time for the remainder of the phase. The total is still 4.5 FTEs, but it is smoothed to enable one person to be assigned.

To the extent that the project manager has followed the resource requirements planning techniques described in the previous steps, the rounding and adjustments from requirements to needs may be relatively small. Nevertheless, this process enables the project manager to use her judgment in the final translation of requirements into needs.

When she completed this planning, Anne was very pleased. "I feel really good that the process of estimating resource requirements and then translating them into rounded needs produced a realistic picture of resource needs on the project. It gives me the resources to complete this phase of the project, and the plan is workable for the resource managers."

Anne also thought that there were some significant cost benefits from the project. "Before I started this new process, using a software tool that is fairly new, I did a resource plan the old "back of the envelope" way. The difference is interesting. The new resource plan has 6.5 fewer FTEs, approximately a 15 percent lower cost. More importantly, the new resource plan gives me a more realistic mix and timing of resources on the project than I would have had without this tool to help me plan."

RESOURCE REQUIREMENTS MANAGEMENT

Resource requirements *management* is the downward component (or right-hand side) of Level 2, illustrated in Figure 7–1, as opposed to resource requirements *planning*, which is the upward (or left-hand side) of Level 2. In resource requirements management, the project manager now has the resources assigned to her project, but they are assigned for a time period. It would be very useful to manage them by step instead.

What advantage would this provide? With resources assigned by step within a phase, the project manager and the team can better allocate their time and measure their time spent against this allocation. Managing resources assigned by step within a project phase can provide some early warning of problems, enabling the team to make the necessary adjustments in time. It also provides an opportunity for the project manager to allocate some buffer time as a contingency, using techniques such as critical-chain resource management to do so.

The top-down aspect of resource requirements management starts with the project manager and team members distributing project assignments for each person to specific project steps.

The top-down aspect of resource requirements management starts with the project manager and team members distributing project assignments for each person to specific project steps. Anne Miller's assignment distribution for Richard Salisbury is illustrated in Figure 7–8. Richard is assigned as 1.0 FTE (full time) for the first two months, 0.5 FTE (half time) for three months, and 1.0 FTE for June. There are five steps in this phase, but

F I G U R E 7–8

Illustration of Assignment Distribution

Richard Salisbury—Marketing

		Phase 1 Planning Step-Level Assignments							Comparisons to Aid Planning	
		Jan	Feb	Mar	Apr	May	June	Total	Std.	Pln.
	Work Days	21	20	22	21	20	20			
	Project Assignment (FTEs)	1.00	1.00	0.50	0.50	0.50	1.00		PD	PD
Step Assignments:										
Project Management	Percentage	25%	25%	25%	25%	25%	25%			
	Person-Days	5.25	5.00	2.75	2.63	2.50	5.00	23.13	30	25
Market Study	Percentage	75%	50%	50%	0%	0%	0%			
	Person-Days	15.75	10.00	5.50	0.00	0.00	0.00	31.25	45	40
Business Plan	Percentage	0%	25%	25%	50%	50%	50%			
	Person-Days	0.00	5.00	2.75	5.25	5.00	10.00	28.00	30	25
Functional Specification	Percentage	0%	0%	0%	0%	0%	0%			
	Person-Days	0.00	0.00	0.00	0.00	0.00	0.00	0.00		
High-Level Software Design	Percentage	0%	0%	0%	0%	0%	0%			
	Person-Days	0.00	0.00	0.00	0.00	0.00	0.00	0.00		
High-Level Mechanical Design	Percentage	0%	0%	0%	0%	0%	0%			
	Person-Days	0.00	0.00	0.00	0.00	0.00	0.00	0.00		
Total Assigned	% of Assignments	100%	100%	100%	75%	75%	75%			
	Person-Days	21.00	20.00	11.00	7.88	7.50	15.00	82.38	105	90
Avail./Contingency (Overassigned)	Person-Days	0.00	0.00	0.00	2.63	2.50	5.00	10.13		

he is only working on three of them: Project Management, the Market Study, and the Business Plan.

Assignment distribution requires a person-day perspective rather than an FTE perspective, so we're back to more granular scheduling. Since the number of workdays is known at this point, the project manager can use this number to be a little more accurate in distributing the time it will take each resource to complete the work. In this case there are 21 work days in January and 20 in February. Anne Miller uses percentages to distribute Richard's assignment. In January, 25 percent of his assignment is distributed to the Project Management step and 75 percent to the Market Study step. For the complete Market Study step, Richard is assigned for 23.13 person-days to Project Management and 31.25 person-days to the Market Study.

Anne uses some comparisons to guide this distribution. She references the estimates that were done in resource requirements planning and the standards established for this resource skill from earlier step guidelines. For example, she assigned Richard to the business plan for 28 days compared to the previous resource requirements planning estimate of 25 and the standard from the step guidelines of 30. There is a significant difference in the market study assignment. It totals 31.25 instead of the 45 person-day standard and the earlier planning estimate of 40. Anne checks with Richard, and he is comfortable that he and the marketing analyst can complete the work necessary.

At this point the project manager can also allocate some of the assigned time as a "contingency" for the entire phase. Doing this is a matter of project management preference and technique. Creating a contingency for the entire phase instead of each step provides a more flexible resource buffer, consistent with critical-chain techniques, when the project manager is using the phase as the resource buffer. In this example, Anne distributed 10.13 person days of Richard's assigned time (a little over 10 percent) as a contingency and put all of this buffer in the last three months, where it is expected to be most useful. Essentially, she has created a project resource buffer for Phase 1 to dampen the accumulated effects of all the uncertainties in all of the Phase 1 steps. Since this buffer is included in the project budget, it will be a favorable variance if not used.

There are some differences between the person-day estimates made in resource requirements planning and the assignment distribution made in resource requirements management. The planning estimates are transformed from person-days to FTEs and rounded to create needs, changing the equivalent person-days. When assignments are actually made, the needs may be altered further in total and in timing to fit resource availability. Finally, at this point, the specific assigned person is known, and the person's skills and capabilities are likely to be different from the generic skills that were originally stated during the "needs" exercise.

In distributing assignments, the project manager may want to take advantage of the specific skills and abilities of the team that has been assigned. She may want to balance actual assignments with steps a little differently based on the final assignments and skills. It may be helpful to distribute an individual's assignment a little differently to the steps he is working on and shift the individual's workload a little across the time period. Or the project manager may want to shift actual step assignments from one team member to another to take advantage of skills or offset skill deficiencies in the actual assignments. Finally, the project manager can make a judgment to allocate some time to a contingency or buffer for the phase.

WORKLOAD RECONCILIATION

The final component in Stage 3, which we introduced in Figure 7–1, is work-load reconciliation. This is the process of comparing work estimates from detailed task scheduling to resource assignments by step. Work estimates are typically focused on the short term: *This is my schedule over the next two weeks for doing the tasks necessary to complete the work needed for the step.* Workload scheduling can be a useful work-management tool, because it can effectively coordinate work among everyone on the team: *I will review the draft specification next Tuesday after you complete it on Monday.*

Detailed task scheduling can also be helpful in making real-time adjustments to work-time allocation if something is delayed in the short term. Some developers can successfully apply this disciplined approach to time management, but others prefer to be less detailed or less disciplined. It depends on the complexity and risk of the work and the capability of the project manager.

When a developer has completed a detailed task plan, it can be useful to reconcile this work estimate against the step assignment to see if there are any differences between the two that require adjustment. Adjustments can be made either by changing the step assignments or by changing the work estimates to fit the assignment allocation. Or the comparison can be useful simply as an understanding of the differences.

This reconciliation of the work estimate with the step assignment can be performed for each individual for each step. In the example illustrated in Figure 7–9, Richard Salisbury's assignment for the Market Study step is 15.75 person-days in January and a total of 31.25 person-days for the entire step. The workload estimate from the task detail shown is 88 hours (11 person-days) in January and 242 hours (30.25 person-days) total. The workload total is less than the assignment by 4.57 days in January, but more than the assigned time by 2.13 days in February and 1.63 days in March. This reconciliation tells Richard that he needs to accelerate some of his work in January. Otherwise, he won't have enough time to complete it by the end of March. So he adjusts the workplan to do task 2.1.4, review purchased market research in January and accelerate the work on task 2.4.2, review potential price ranges. Richard finds this to be very helpful. "Previously, I would have followed the detailed workplan and run out of time in March, possibly delaying the project. With my assigned time distributed by step, I'm able to coordinate my work better within the assigned time."

Generally, the usefulness of this reconciliation exercise depends on the individual. It may not be worthwhile for some team members who don't want to do detailed work scheduling, so it's best applied only when it will be utilized.

F I G U R E   7–9

Workload Reconciliation Example

Richard Salisbury

	Jan 21	Feb 20	Mar 22	Apr 21	May 20	June 20	Total
			Phase 1 Planning				
			Step-Level Assignments				
Step Assignment (Person-Days)	15.75	10.00	5.50	0.00	0.00	0.00	31.25
Workload Total from Detail	11.00	12.13	7.13	0.00	0.00	0.00	30.25
Variance	4.75	-2.13	-1.63	0.00	0.00	0.00	1.00
Detail Task Work Estimates from High-Level Tasks (Hours)							
2.1.1 Identify available market data	12						12
2.1.3 Define product opportunity	32	20					52
2.1.4 Review purchased research	0	32					32
2.2.5 Review draft customer questionnaire	20						20
2.3.2 Identify target customer	4						4
2.4.2 Review potential price ranges		5	5				10
2.7.3 Review preliminary market size estimates		20					20
2.9.3 Summarize major customer benefits	20	20	20				60
2.12.5 Review and finalize			32				32
Total Hours	88	97	57	0	0	0	242

Done this way, resource requirements planning is not dependent on the acceptance of disciplined workload management by all team members.

Some people also find it useful to use a bottom-up detailed task schedule to estimate resource requirements for resource requirements planning. We illustrated this in Figure 7–4, with the use of a workplan estimate for resource requirements planning.

However, there is a real danger in bottom-up resource planning. Resource management done solely bottom-up can unnecessarily lengthen time to market and resource requirements for many projects. If detailed work estimates at the task level are rolled up to determine project schedules, then project schedules increase to accommodate that time. For example, a marketing manager does a detailed task workplan for developing a business plan, and the roll-up of the detail indicates that it will take him 280 hours of work. The three others working on the business plan estimate their workloads at 220 hours, 230 hours, and 170 hours, respectively. Bottom-up project planning indicates that the business plan step will require seven weeks to accommodate the 280

Resource management done solely bottom up can unnecessarily lengthen time to market and resource requirements for many projects.

hours of work, instead of the six weeks that it should take. In reality, the team will find a way to complete the step in six weeks by redefining the work, reallocating some of the work, or working more than 40 hours in some weeks.

We reviewed this precept in our example of the electrical design engineer giving Anne Miller an estimate of 55 person-days to do the business plan instead of the 30 person-days that the step library guidelines indicate are standard. Anne reviewed the workplan but decided to go with 25 person-days instead, because much of the work defined didn't need to be done in this phase. The 55 person-days would have unnecessarily lengthened the project time.

Bottom-up planning requires an institutional commitment to disciplined and detailed work scheduling. Everyone must be comfortable that detailed task planning can be done, even for project steps that may lie a year or more in the future. It also requires a level of standardization at the detailed task level, which is much more difficult to achieve than it is at the step level, in order to create resource requirements from detailed tasks for later phases of a project. All of these issues call into question the usefulness of detailed task planning as the basis for bottom-up resource requirements planning.

REQUIREMENTS

Stage 3 requirements build from those of Stages 1 and 2. The difference in this stage is that it can be implemented project by project, so adoption can be more gradual. Project managers can decide for themselves if and when they find these Stage 3 resource requirements planning tools helpful.

Stage 3 systems requirements are essentially a set of tools for the project manager to use in doing the resource requirements planning and management that sit on top of a resource management system, which is part of an integrated DCM system. A resource management system includes:

  ◆ Resource requirements planning tools that enable the project manager to make resource estimates by skill category for each step. This may include the ability to reference workplan estimates and standards.

These tools may also automatically incorporate resource estimates from a step library if one is used.

♦ Resource balancing tools that enable the project manager to time-phase steps and shift work among resources to balance resource requirements and fit them to available resources.

♦ Tools for translating resource requirements to resource needs, which help the project manager easily make this translation and round resource needs prior to scheduling them.

♦ Tools for distributing assignments to steps, which enable the project manager to manage resources at the step level and create phase resource buffers.

♦ Tools for reconciling workload estimates to assignments, thereby helping developers who want to perform this reconciliation.

While this can be done to some extent using spreadsheets, the results need to be manually transferred, and some of the value of integration will be lost. It is much easier for project managers to use an integrated tool set. In either case, the resource management system needs to support the concepts described in the full range of capabilities in Stages 1 to 3.

There are few broad process requirements for Stage 3, since the improvements are at the project level. It is mostly a matter of acquiring the DCM software tools, making them available to project managers, and training them in how to use them effectively. Stage 3 does require, however, that the concepts established in Stages 1 and 2 be fully implemented as the standard across the company, since Stage 3 does manage the translation among resource management levels. Some companies may want to more aggressively go after the benefits from this new stage by establishing these practices as standards and measuring the improvements in their resource requirements planning.

BENEFITS

As we discussed at the outset of this chapter, the objective of Stage 3 is to optimize the resource assignments to individual projects. The benefit is a reduction in project development cost, which can come from several sources. Most companies will be able to see a cost reduction of 10 to 25 percent in most projects by implementing Stage 3 capabilities. Additionally, in some cases better resource planning will reduce the number of project delays.

1. Through resource requirements planning, the project manager can better estimate required resources at the step and skill level.

2. Through resource balancing, the project manager will be able to better balance resource estimates and eliminate unneeded resources.

3. Through the use of standards or estimates from step libraries, the project manager can easily leverage accumulated organizational experience.

4. Through a process of translating resource estimates to project needs, the project manager can better fit resources to the project and reduce excess resource needs.

5. By translating actual assignments to project steps, the project manager can better manage progress of the project.

6. By using resource buffers for a phase, a project manager can reduce project risk due to resource constraints.

7. By reconciling workload estimates to project assignments at the step level, developers can better manage their time.

SUMMARY

In Stage 3 we shifted the focus from managing resource utilization across all projects to managing resource productivity in individual projects by more efficient resource planning. We introduced four new practices for achieving this: project resource requirements planning, translating resource requirements into resource needs, project resource requirements management, and workload reconciliation. We described these practices in detail and illustrated them with examples of how Anne Miller applied them to the Fast-Food Robot project of our hypothetical company.

8

CHAPTER

Stage 4—Fully Integrated Resource Management

Stages 1 through 3 achieved the majority of the utilization and productivity benefits of resource management—perhaps as much as 85 to 90 percent, but some benefits still remain to be captured. These have to do with completing the integration of resource management systems and processes with other systems and processes. This integration not only reduces the administrative time and risk associated with partial or manual integration, but also makes the previous benefits of improved resource management more sustainable. This final stage of resource management maturity focuses on seamlessly and automatically integrating resource management with other internal systems and processes, as well as with external resource systems, in order to extend the capabilities of Stages 2 and 3. Although Stage 4 is the final stage, some companies may find it beneficial to do some of the integration as they implement previous stages, and then complete the integration in Stage 4.

The benefits of Stage 4 come from this seamless integration. They include the efficiency gains from eliminating duplicate entry of information from system to system or process to process. They also include the efficiency gains that come from eliminating the delays and potential for errors that result from transferring information between systems and across processes.

More than even the previous three stages, the implementation of Stage 4 capabilities will vary from company to company, depending on the other processes and systems the company has in place and its priority for systems integration. So, rather than providing a specific roadmap for integration in this chapter, we will review a general overview of the major points of resource management integration.

The points of resource management integration can be organized by level in our three-level resource management framework, which was introduced in Chapter 4 and is illustrated in Figure 8–1 with the integration emphasis added. Describing these points of integration by resource management level helps to indicate the type of information that is integrated and the type of integration made possible, since the characteristics of data vary at each level.

Mahesh Gupta, the CIO at Commercial Robotics Inc. (our hypothetical company), was responsible for planning the integration of the company's new resource management system with other systems at CRI. Mahesh explained his general philosophy of integration this way: "There are three different types of systems integration, and it's important to understand the differences. The first is easy integration, where one web-based application simply links the user to another. The second is integration that requires the transfer of data between two applications or access by multiple applications to the same data that are used for different purposes. That is more difficult, but is possible. The third is

F I G U R E 8–1

Resource Management Points of Integration

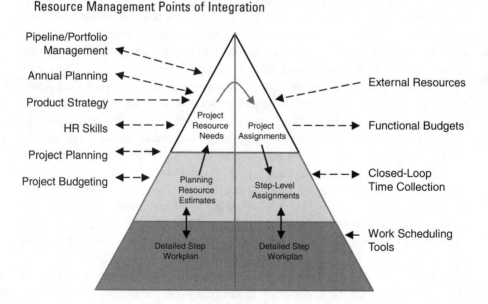

where different applications need to work together to perform an integrated function. Here, our preference is to purchase an integrated system from a single vendor, because this type of integration is very difficult and causes continuing maintenance problems."

LEVEL 1 POINTS OF INTEGRATION

At Level 1, resource scheduling information—project assignments and needs—are the primary basis for integration. Typically, these assignments and needs are FTEs assigned to projects, as well as to nonproject activities.

Integration with Pipeline/Portfolio Management

Resource management is usually tightly integrated with pipeline and portfolio management so that project portfolio optimization can be balanced against capacity. Without including this resource constraint, portfolio decisions are quite limited. Due to the importance of tight integration, resource management systems and pipeline/portfolio management systems are almost always part of a common DCM system. We'll look at this integration in Chapter 18.

> Resource management is usually tightly integrated with pipeline and portfolio management so that project portfolio optimization can be balanced against capacity. Without including this resource constraint, portfolio decisions are quite limited.

These systems need to be tightly integrated for several reasons. First, the data formats used by each system need to be the same. The systems have to define assignment data the same way, use resource needs in a similar manner, follow the same phase-review process, apply the same skill categories, and group resources by the same categories and resource groups. Translating resource and capacity information from a resource management application to a very different portfolio management application can be difficult, if not impossible. Manually transferring the information can be time-consuming, and developing an automated interface can be very expensive and unreliable. CRI used a common development chain management system because this integration was built in.

Reconciliation with Functional Budgets

The achievement of Stage 1 resource management capability enables a company to reconcile resource assignments with functional budgets. This reconciliation is a major misalignment in most companies. All companies establish annual functional budgets that include the cost of resources in all resource groups. They also approve project investments based on proposed project budgets throughout the year. While the real investment decisions are made in the annual budgeting process, these decisions are not reconciled with the strategic decisions supposedly made by allocating resources to projects.

This reconciliation can be accomplished through a charge-out process, where every resource group is measured as a cost center with the expectation that it will break even by "charging" resources to approved product development projects. The integration of resource management comes from providing the information on project assignments to the financial application that does the reconciliation between the functional budget/actual costs and the actual "charge out" from assignments to approved projects.

At the end of each month, CRI's resource management system transferred assignments in FTEs, and the related resource costs, to its ERP software. These data were then used to create a financial report for each resource group that showed how much of its costs were "charged out" to projects through assignments. We'll look at how this is done in Chapter 19.

Integration with Annual Financial Planning

As you may recall, we discussed the integration of medium-term resource planning with annual financial planning in Chapter 6. By extending the horizon of projects included in resource management, resource capacity planning can be extended through the full annual financial planning horizon, and then project resource needs will include resource needs for the upcoming year. These resource needs can then be used to prepare the resource group headcount plans for the annual budget. This will also be examined in Chapter 19.

This is more of a process integration than a systems integration, because resource group managers can easily use the medium-term capacity planning information from resource management to prepare their budgets. There is little value in integrating the resource management systems and financial planning systems. What is more important is the integration of resource management systems with portfolio/pipeline management, since it's likely that the assumptions regarding which projects will be started next year will need to be adjusted to fit with realistic resource capacity plans.

At CRI, even though the integration of medium-term resource planning with annual financial planning was a major objective, Mahesh found that systems integration wasn't needed. Resource management tools provided resource capacity planning reports and charts to the resource group managers for them to use as input in their planning. They could easily see how many people they needed, and make the translation manually into what headcount they needed to plan for. The real work came in cutting back the portfolio to fit the fiscal constraints of the R&D budget.

Integration with Project Budgets

Resource costs are typically the most significant of all development costs, so integration of resource planning and management systems with project budgets is essential. A fully integrated project budget changes to reflect changes in resource assignments so that the resource plan and project budget are always aligned.

Project budgets include the costs of project resources in various forms. Assignments are included, based on the specific person assigned, typically using a standard cost for that person. Resource needs are used for future project phases; their costs are computed based on a standard cost for the skill. In some cases, a company may want to break out costs by project step and compute the resource costs based on step-level resource assignments.

As a project progresses, portions of the project budget become actual rather than planned, and the budget should reflect these actual costs. This involves not only recording the actual cost based on actual time, but also adjusting the budgeting going forward based on what is required to complete the project. Eventually, the entire budget becomes an actual cost. We'll go into more detail on integrated project budgeting in Chapter 13.

Product Strategy Integration

As we discussed, medium-term resource capacity planning is improved when the time horizon for resource needs is extended beyond current projects. But doing this requires the integration of formal product line planning. Some DCM systems include planned products as projected resource needs for purposes of resource capacity planning. These projected needs can be either included manually or integrated automatically through a common systems interface between the product strategy system and the resource management system.

As part of its DCM system, CRI implemented a new product strategy system that provided a set of tools to the product planning teams to help them develop product line plans and identify new product platforms. One of the most important outputs was a master product line plan that identifies the product roadmap for the company. This application was used by CRI to estimate resource needs for the products on its roadmap, which was automatically converted into a format that could be incorporated into medium-term resource capacity planning in the resource management system. We'll look at product strategy and planned products in Chapter 20.

Integration with HR Systems Skill Categories

Most companies maintain certain resource information, such as skills for all employees, in their HR or ERP systems, and would like to use the same information as the basis for specifying the skills of individual resources and the skill categories for needs during resource planning. This would eliminate the cost of duplicate data entry. In some cases this integration may be easy, but in others it may be more difficult if the HR and resource management systems use the data differently.

CRI found it difficult to integrate data from its HR system into its resource management system. Instead, it decided to develop common, front-end software that enabled one-time data entry and provided the necessary data to each system.

Integration with Multiple Resource Groups

Most multidivision companies will manage resources and maintain resource management systems by division, since this is a primary responsibility for each division. There are times, however, when it's a very good idea to share resources across divisions, and this can be done by integrating resource management systems across the company.

This system integration involves giving each division limited access to the resource and availability information of other divisions. It also calls for defining the management process for assigning resources across divisions, for determining the practices for transferring costs, and for deciding to what extent resource needs, requests, and assignment capabilities will be shared. Although there will be resistance to this resource sharing across divisions initially, I expect that it will eventually become a standard practice.

Integration with External Resources

In the future, the most exciting aspect of resource management integration may well be the integration of external resources. When a company leverages codevelopment and does more outsourcing, it may want to directly integrate the resources of outside resource groups into its resource management systems and process. A company can identify resources available from outside groups when it does assignments and even when doing resource requirements planning. For example, if a specific skill is required that is not available within the company, the project manager could tap into an external resource group to find a consultant or contractor who would be available to do the work.

This external resource integration would include an integrated resource transaction process, giving a project manager the ability to request specific resource assignments from outside resource groups. This external resource integration is similar to the integration of resource groups from other divisions within the same company, but contains some additional access control. Eventually, outside contractors will provide this integration as a competitive differentiator.

> In the future, the most exciting aspect of resource management integration may well be the integration of external resources.

CRI integrated its resource management with several of its codevelopers as well as a number of subcontractors. Each of these external resource groups gave CRI access to its resource availability through an integrated external resource system. Each codeveloper was established as an external resource group, and it maintained the resource availability information on all of its people. When CRI scheduled one of these resources on a project, that assignment was indicated on the shared resource system, and the resource was treated as one of CRI's normally managed assignments. External resource group information was screened so that CRI only had visibility of availability and assignments on its own project, not assignments in support of other companies. Subcontractors were grouped by general skill-set, and they periodically updated CRI on their availability. CRI managed these resources as another resource group, but excluded them from any utilization reporting.

Incorporating external resource groups into its resource management increased CRI's flexibility and responsiveness. CRI's Anne Miller, manager of the Fast-Food Robot project, shared her thoughts about this flexibility gain. "We project managers can now turn to external resources quickly when we don't have the internal resources. It's almost as easy as doing it internally; we just need to identify and justify this 'need' at our phase review. When we see that we don't have the necessary skills available internally, we simply access these external resource groups to see if someone is available with the right skills. If they are, we then establish a need for this resource, and that automatically reserves the person for 10 days until we get the approval. I recently used this to replace someone on a project who had to go on a leave of absence, and it took me only 15 minutes to locate the right person. He started on the project two days later. Previously, this would have taken 6 to 10 weeks."

LEVEL 2 POINTS OF INTEGRATION

Generally, the points of integration in Level 2 require tight integration and are almost always provided by the same integrated development chain management system. This is what CRI did, based on the previously discussed integration strategy of its CIO: It used one system.

Integration with Project Planning

The resource management system must be tightly integrated with the project planning system. Resource estimates, needs, and assignments should be based on the project steps; when these steps are rescheduled, the resources should be automatically rescheduled. Because of this need for tight integration, project planning and resource management applications usually reside in the same integrated development chain management system. This also requires what I call an enterprise project planning and control system, as we will see in Chapter 10.

A company may choose to use project work-scheduling tools such as Microsoft Project for workplans and workload estimating, but these would also be integrated at Level 3.

Closed-Loop Time Collection

Simply collecting the actual time spent on projects is of limited use without being integrated into a closed-loop system. Knowing that someone spent 15 person-days last month on a project can be useful in tracking what the person is actually doing, and, in some companies, this type of information provides the only visibility into utilization. But this is historical information, and cannot be used proactively to improve utilization, optimize resource allocation, or avoid bottlenecks.

What is more useful is a closed-loop time-collection system for comparing what developers are actually doing to what they were assigned to do. It's helpful to understand this correlation or lack of correlation in planning. The information can also be used to initiate corrective action to bring actual efforts in line with plan.

Such a time-collection system can also be used by project managers to obtain visibility into progress and percentage of completion of specific project steps. For example, a developer is assigned to work 10 days per month over three months to complete a step, but actually spends 30 days over the first two months. It could be that the developer completed his work on the step early, or perhaps the work is taking much longer than planned. A closed-loop time-collection system can also collect status or percentage-of-completion information from the developer to complete the project manager's visibility of how time is being used on a project.

It's best to compare actual to assigned time at the step level. This is more informative than collecting it at the phase level, particularly if the project manager wants to get the visibility needed to make adjustments, and it is more meaningful than collecting it at the task level, which is much too detailed to be useful. It takes too much time for developers to record their time across dozens or hundreds of tasks, and, for that reason alone, whatever they record is usually inaccurate. It is also difficult for a project manager to make any worthwhile judgments at a very detailed level.

CRI implemented a closed-loop time-collection system in its implementation of Stage 4 capabilities. The time collection for a typical week is illustrated in Figure 8–2 for Richard Salisbury's time. His time-entry form shows

F I G U R E 8–2

Closed-Loop Time Collection

Richard Salisbury	Actual Time							Week	Schd.	Diff.	Proj. Adj.
	M	T	W	T	F	S	S				
Fast-Food Robot Project											
Project Management	4				2			6	10	-4	-4
Market Study		8	8	8				24	20	4	0
Business Plan					6	6		12	10	2	0
Project Total	4	8	8	8	8	6	0	42	40	2	
Nonproject Time											
Training								0			
Customer Support	4			2				6			
Vacation											
Holiday											
Sick Time											
Other											
Total	8	8	8	10	8	6	0	48	40		

his scheduled time for the three steps he was assigned to for the first week in February on the Fast-Food Robot project: 10 hours for project management, 20 hours for the market study, and 10 hours for the business plan. Richard recorded his actual time by day on the time-entry form for each step: four hours on project management on Monday, eight hours on the market study on Tuesday, etc. He also recorded the time he spent on customer support, even though he wasn't scheduled on it.

CRI also collected what they called "quick status" feedback on the time-collection entry form from each developer. As Anne Miller explained, "This is helpful to the project managers because it gives us some insight into what the developers were thinking when they reviewed their time." In this case, Richard spent 6 hours on project management instead of the assigned 10 hours, but he recorded a project adjustment of -4, which was the notation to indicate to the project manager that he completed the project management work for the week and did not need to carry the unused time forward. On the market study he spent 24 hours instead of the 20 planned, and he indicated that the additional four hours was extra and absorbed in the week by indicating a 0 adjustment. Anne explained her interpretation of this: "It means that the extra four hours were incurred, but that it won't reduce the future time assigned to the step."

Integration with Knowledge Management Systems

Resource management may also integrate with a knowledge management or a resource standards system to provide resource guidelines through project templates or step libraries. This integration can be manual or automated. Manual integration is done by entering or referencing resource guidelines outside of the resource management system. Automated integration can be done through systems integration, or (even better) when resource guidelines are incorporated into the same development chain management system. We'll discuss how this works in Chapter 16.

LEVEL 3 POINTS OF INTEGRATION

At Level 3, the integration is of detailed project scheduling tools, such as Microsoft Project, that are used to prepare and maintain workplans. It's best to set up these workplans by project step, so they can be more easily integrated. Integration at this level typically supports the reconciliation that was described in the last chapter, and compares the work estimate computed from the detailed workplan to the step-level assignments.

REQUIREMENTS

Stage 3 requirements typically focus on systems or process integration of the changes implemented in previous stages.

Our assumption in this discussion is that the resource management process is already employing a highly integrated system: Most of the resource management activities described in Stages 1 through 3 use a single integrated DCM system. The exception is the integration of desktop-based, detailed project-scheduling applications such as Microsoft Project. These are sufficiently generic that the interface can be incorporated into the resource management system.

As was discussed at the outset, specific system integration requirements depend on the systems integrated and the priorities of the company.

Process integration requirements also vary widely. Integration with annual financial planning requires synchronization of the medium-term resource capacity planning process and the annual financial plan. Functional budget reconciliation requires a formal process to define how functional managers are expected to manage the reconciliation.

Implementing a closed-loop actual-time collection process requires a significant process definition and widespread training for all developers. Each person needs to know what is expected and how the information is used, so the implementation may take some time.

BENEFITS

The benefits from Stage 4 come from the efficiency gains achieved by eliminating the need for redundant entry of the same or similar data. There are also significant execution benefits from eliminating the delays and errors that result from systems and processes that are not integrated, and this makes the previous benefits more sustainable. Finally, rapid execution and response require seamlessly integrated systems and processes.

SUMMARY

Stage 4 addresses the integration of development resource management systems and processes with the other R&D-related systems and processes. We have examined the points of integration for each level of the three-level framework we've used throughout Section Two of this book.

The system, process, and practice improvements contained in the first three stages of development resource management maturity account for some

85 to 90 percent of the potential gains and benefits. But it's important to emphasize that any company that stops at Stage 3 will end up forfeiting more than the final 10 to 15 percent of the benefits of world-class resource management. They'll soon find that some of the gains they achieved from short-term utilization management (Stage 1), medium-term capacity planning and management (Stage 2), and resource requirements planning (Stage 3) are suffering from erosion. The reason? The advances at each stage are information-driven, and the volume and variety of the information involved preclude it from being driven through the entire development system effectively or sustainably by informal, manual, or disconnected means. System and process integration is indispensable.

Project Management

9
CHAPTER

Stages of Project Management

The Time-to-Market Generation of R&D management focused primarily on project management, and there were many important management advances that resulted in some impressive benefits. Consistent with the theme of that generation, time to market was reduced by 40 to 60 percent in most cases, but there were other benefits as well. As project development time was reduced, developers did not work as long on each project, and this generally reduced the project cost. Because most projects were now following a standard process, they were much more predictable, and unexpected delays that were previously characteristic of product development were greatly diminished. Development waste was greatly curtailed as inappropriate projects were canceled earlier and risk was reduced. And finally, an unexpected benefit—better products—was also achieved at many companies. The new generation of R&D organizations will build upon the project management achievements of this previous generation.

In this section, we'll examine the new generation of project management precepts, practices, and modes of organization. We'll look at the limitations of previous project management practices, and show how the driving philosophy and practices of the new generation will overcome those limitations. But before going into the details, you may find it helpful to see a preview of the changes in store. As in the previous section of this book, we will use a stages-of-maturity model to group the new practices. In the project management stages there have already been some improved practices from the previous generation to build upon.

The stages of capability in project management are defined in Figure 9–1, including the generation corresponding with each stage. (Note that the R&D Productivity Generation spans two stages.) Stage 0, informal project management, shows a lack of much attention to project management. At this stage, project results are haphazard, and today almost all companies have progressed beyond this stage. Stage 1 marks the introduction of some basic project management disciplines, which we discussed in Chapter 2. These disciplines are typically applied within each function separately, hence the term "functionally focused project management" used to describe Stage 1.

The second stage of project management was the primary thrust behind the Time-to-Market Generation. Stage 2 is the stage at which phase-based decision making, cross-functional core teams, and a structured development process are introduced. This stage includes the PACE process for product development, as

F I G U R E 9–1

Stages of Project Management Maturity

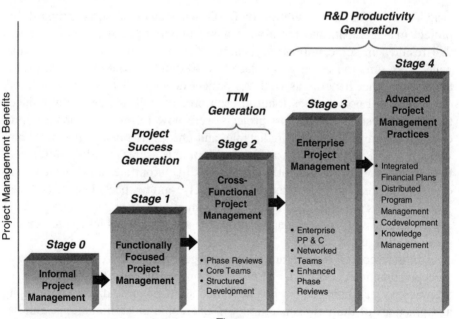

well as similar processes. As we discussed previously, the benefits of this stage, and the TTM Generation it created, are very significant.

The next generation of product development defines project management as incorporating two distinct stages: Stage 3 for enterprise project management and Stage 4 for advanced project management practices. Two stages are helpful here because Stage 3 provides the indispensable foundation for Stage 4.

As you will see, the three primary practices of Stage 2—structured development, cross-functional core teams, and phase reviews—are significantly improved in Stage 3. Structured project planning uses the same structure in Stage 2, but it evolves in Stage 3 from a technique for the project manager to a management process, namely enterprise project planning and control, which is shared by everyone involved with product development. Cross-functional core teams are replaced by networked teams that efficiently overcome the limitations of time and distance. The phase-review process of Stage 2 continues, but is enhanced by the integration of information from enterprise project planning and control.

Stage 4 introduces four new management practices, which, again, are enabled by the foundation laid in Stage 3. The four practices are integrated financial planning and project budgeting, distributed program management, collaborative development, and context-based knowledge management.

STAGE 3—ENTERPRISE PROJECT MANAGEMENT

The three management practices of Stage 2 project management (and of the TTM Generation) palpably improved the development process, but with some limitations. In Stage 3, development chain management (DCM) systems enable companies to overcome these limitations, take advantage of some new opportunities for improvement, and build a foundation for continued improvement. We'll look at these improvements as enhancements to the three Stage 2 practices.

Enterprise Project Planning and Control

Structured development techniques introduced the advantages of top-down project planning in the Time-to-Market Generation. By applying these techniques, product development project plans were created from consistently applied, standard cycle times for project steps. Standard cycle times significantly eliminated the natural tendencies for extending cycle time by rolling-up work estimates, enabled best practices in project management to be applied across all projects, and minimized unexpected project delays by applying proven cycle times to estimates of new project completion dates.

Eventually, many companies encountered limitations to applying these structured development techniques. Since project plans were created and maintained solely by project managers using desktop software tools, there was little integration and collaboration of the project plan. Project planning had to be done without visibility of resource availability. The use of standard cycle times relied on project managers following published guidelines. And project team members couldn't use the project plan as a collaborative system because it wasn't a system in the sense that the plan didn't let them see where the work stood, who was doing what, what had been decided, etc.

In Stage 3, project planning and control tools used solely by project managers are replaced by an integrated project planning and control system, along with the management process to raise planning and control to an entirely new level. This system will enable some important new capabilities, but the most significant implication is that project planning and control will be transformed from a technique to a company-wide management process—hence the use of the term "enterprise project planning and control" to describe it.

We'll also expose a major trap that awaits companies that abandon a top-down project planning process for one that is bottom-up. Companies that do this unwittingly, in order to accommodate the requirements of some DCM systems, will end up losing many of the benefits gained in the TTM Generation without knowing it until it's too late.

Networked Project Teams

One of the most significant organizational improvements of the Time-to-Market Generation was the creation of cross-functional core teams. This new model of the most basic organization in product development—the project team—clearly focused and simplified communications, coordination, and decision making. Its early limitation was that to be highly effective it required core team members to be colocated, and to be dedicated to a single project.

Companies tried using email to overcome the colocation restrictions, but I believe that this compromised the core team organizational model by bypassing some of its important communication and coordinating mechanisms. The solution, of course, is not to eliminate email, but to use the information technology of DCM systems to create a more advanced organizational model: the networked project team.

We'll examine how the networked project team is superior to the cross-functional core team model of the TTM Generation, and how it corrects the compromised project communications and coordination inherent in email.

Enhanced Phase-Review Process

Phase-based decision making was perhaps the most important innovation of the TTM Generation. It enabled companies to invest in new products on a phase-by-phase basis, and to cancel weak projects much earlier. Some companies saved as much as 10 percent of their R&D budgets simply by canceling their investments in inappropriate projects sooner rather than later.

A phase-based decision process requires approvals of a project at each phase of its development, based on specific criteria for completion of the previous phase and preparedness to enter the next phase. In most companies, the authority to make these decisions was invested in a new organizational entity, generically referred to in PACE terminology as the Product Approval Committee (PAC). This instituted a new development management process as well: the phase-review process.

Development chain management systems provide some new capabilities for the PAC and expand the capabilities of the phase-review process. These systems make resource assignment information available to project managers and the PAC, giving them visibility of resource availability for their decision making. In many companies this visibility to more information will change the phase-review process from isolated decisions on individual projects to comprehensive decisions on groups of projects, with an explicit focus on how resources should be allocated. As we will see, companies that combine this resource-conscious decision making with improvements in the strategic aspect of product development may want their PACs to shift their emphasis from strategy setting to strategy execution.

> Visibility to more information [through DCM systems] will change the phase-review process from isolated decisions on individual projects to comprehensive decisions on groups of projects.

DCM systems will also provide the Product Approval Committee with its own information system, enabling it to better manage its work as a project review and approval body. This information, while a by-product of the other DCM systems, can be very helpful to the PAC. Another important improvement comes from the capability of the DCM system to automatically monitor changes to specified project tolerances, sometimes referred to as project contract items or boundary conditions; and to automatically inform the project team and even the PAC when a project exceeds these boundary conditions. We'll get into more detail on these Stage 3 practices in the upcoming chapters.

STAGE 4—ADVANCED PROJECT MANAGEMENT PRACTICES

Once the foundation of enterprise-level project management is established in Stage 3, certain more advanced project management practices can be introduced. It's not necessary for a company to implement these practices in any particular sequence, and, in fact, it may choose to implement these practices selectively, depending on its priorities and strategy. We'll summarize these practices now, and then explain them in more detail later.

Integrated Financial Planning and Project Budgeting

The expected success of a new product is expressed in financial terms in new product financial planning, and the cost of developing this new product is expressed in financial terms in the project budget. Today, most companies do this planning and budgeting using individual spreadsheets located on the project manager's or financial team member's computer. Consolidating critical financial information across projects is very difficult, if not impossible, when spreadsheets are used in this way. In some cases, the spreadsheets are inconsistent across all projects; and all too often they are financially flawed, leading to incorrect project cost and benefit assessments. Finally, while financial information is really derived from other operational information, this derivation is rarely automatic, making the synchronization of financial and operational information very difficult. This is where the information-processing automation afforded by DCM systems is useful.

Increasingly, chief financial officers are realizing that they are responsible for controlling this critical financial information, and are implementing more integrated new product financial planning and project budgeting to overcome these internal control weaknesses. In this new generation of product development, this integrated planning and budgeting will be accomplished through an integrated system and process as part of a broader DCM system. These capabilities will enable the CFO to implement standard financial planning and budgeting models that are consistently followed by all project teams, to rely on the financial information derived from the underlying operational data and standard financial assumptions, and to put in place an integrated system for consolidating financial information across all projects.

Distributed Program Management

Product development projects are not always independent of one another. Two projects, and sometimes more, may be in simultaneous development at different stages for different sequential versions of the same product. Multiple new products could be in development that are based on the same product platform and share some underlying technology. The development of a new technology, and the development of the product that will incorporate the technology, may be conducted simultaneously as separate projects. Other development projects might not have a product focus at all: for example, a project to develop the new manufacturing process or facility for a new product. And some projects might be so complex that they are better broken up into multiple projects to make them more manageable. The management of interdependent projects such as these is referred to as distributed program management.

Distributed program management is enabled by a DCM enterprise planning and control system, which provides the platform for the integration of related projects within a development program. You will see in Chapter 14 that there are three dimensions to distributed program management: integration of planning and control among multiple projects; integration of the financial information of related projects to provide a consolidated program view; and collaboration of developers working on these related projects, using what I refer to as a network of networked teams.

Collaborative Development Management

Product development is increasingly performed in collaboration with external partners, customers, suppliers, and contractors, as companies try to reduce the cost of development and make development costs more variable. But progress on codevelopment has been slow because the enabling technology and management processes were missing. After all, how can project teams work collaboratively with external partners if they don't even have the systems for intra-team collaboration?

DCM systems provide a codevelopment platform as an extension of the enterprise project planning and control process. Codevelopers will be able to collaborate seamlessly with the project team, while at the same time maintaining the confidentiality of project information. This is similar to what the network of networked teams does.

Context-Based Knowledge Management

R&D is all about knowledge, and you would think that knowledge management is well advanced in all R&D organizations, but it is not. The primary constraint is the inability to deliver the necessary knowledge to the right person at the right time. Important best-practices guidelines, technical data, and reference materials may be available, but developers typically find that locating such information is frustrating and time-consuming.

The DCM systems and processes of the next generation of development management will provide the platform to deliver critical knowledge to developers who need it, when they need it. I refer to this information delivery as just-in-time or context-based knowledge management, which we'll examine in more detail in the final chapter of this section.

SUMMARY

In summary, this chapter shows how Stages 3 and 4 of project management maturity can be achieved through enterprise-wide project management and advanced project management practices and systems. These practices and systems enable the following capabilities: integrated financial planning and project budgeting, distributed program management, collaborative development, and context-based knowledge management. In Stage 3, cross-functional core teams are replaced by networked teams, and new systems allow greater visibility of more project information to more layers of management, allowing comprehensive decisions on groups of projects, not just isolated decisions on individual projects. One of the distinctive benefits of Stage 4 will be that R&D executives and financial managers will finally have the necessary perspective on the entire project portfolio to make fact-based decisions on R&D investments and resourcing.

10
CHAPTER

Enterprise Project Planning and Control Process

In the previous product development generation (the Time-to-Market Generation), structured development was introduced and used to define a hierarchy of project phases, steps, and tasks, enabling companies to apply top-down cycle-time standards to project planning. This new discipline reduced cycle times and greatly improved the reliability of project plans. In most cases, the top levels of this structure—phases and steps—were accomplished through manual techniques, such as project overview charts. While effective, there were obvious limitations to these manual practices, including preparation time, keeping them accurate, and integrating them across projects and within projects.

Project managers used software tools to plan and control their projects, but generally the tools ran on the project manager's (or her assistant's) computer, and were used exclusively by the project manager. She would prepare the project plan, make copies for everyone on the team or distribute copies electronically, and periodically review the plan in meetings, only to find at some point that it was out of date, at which point she would go back to her office to update the plan.

Eventually, the limitations of these software tools became apparent. First, the project plan was all too frequently considered out of date because the project manager was the only one responsible for keeping it current. Sharing the responsibility for the plan was difficult, since the plan resided on one person's computer and only that person could maintain it. Second, the plan was passive and not interactive; it was static, and couldn't be used as an active framework for the project, so team members were unable to collaboratively use the project plan to control the project as it progressed. Third, because project planning was an isolated tool, resource-based project planning was virtually impossible, and project plans had to assume infinite resource availability. Finally, the standardization of project planning around best practices was elusive because it relied on published standards instead of on an integrated system.

With DCM systems, project planning and control tools are now being replaced by enterprise-based applications: systems that extend project planning and control beyond the project manager. With the integrated capabilities of DCM systems, the entire project team actively participates in project planning and control, transforming it from a management *technique* to a management *process*. This is a very exciting change! Everyone working on the project can now use the project plan to see what is really going on, immediately communicate status and changes, and partake in a common view of the project. We will refer to this advancement as an *enterprise project planning and control process* in order to distinguish it from project planning and control techniques.

The transition of project planning and control from a technique used by an individual to a broad-based management process is the central theme of this chapter. We will not attempt a broad review of project planning and control techniques, since there is already a lot written about these. Instead, we will focus on the key aspects of enterprise project planning and control, and the transformation of project planning into a *process* through the following changes:

- ◆ The project plan will no longer be just a document, but an enterprise-wide system for project planning and control, centrally located so that everyone has access to it.
- ◆ Under an enterprise project planning and control system, the project plan will have many active users, not just the project manager. This wider use will introduce the opportunity for *collaborative* project planning and control.
- ◆ With an enterprise system, instead of an individual tool, companies will be able to establish common development practices that are clearly understood and consistently applied across the entire organization.
- ◆ Project planning will be highly integrated with resource management, enabling resource-based project planning.

We will go into more detail on the enterprise project planning and control process, including how it is used for collaborative project planning and control, as well as how it integrates with resource availability and project planning standards. We will also explain how an active enterprise project plan establishes the foundation for the networked project team, provides the basis for context-based knowledge management, and enables distributed program management. First, however, we will examine the importance of enterprise project planning architecture.

ENTERPRISE PROJECT PLANNING ARCHITECTURE

When project management was a technique based on project management software tools, project planning and management were elements of a free-form exercise. Some did it well, while others had difficulty, but it was left to individual project managers to do it their own way, for better or for worse. Now that project management is being established as a common process across the entire enterprise, it's important to define a process that reflects best practice, and this begins with the definition of a process-centered architecture for project management.

Management process architecture is a very important, but frequently underrecognized principle. It is the way a company defines its *management* process, and it, in turn, enables or constrains management best practices. A good management process architecture improves performance, while a bad one inhibits performance, even by capable people. Yet companies all too often ignore this principle. They acquire application software, and let the architecture of the software dictate their management process architecture. Unfortunately, application software doesn't always enable management best practices. Sometimes the software is implemented incorrectly or incompletely, with mediocre results. (This was the reason behind much of the criticism of ERP systems in the 1990s.) In other cases, software may not be designed around best practices, in which case it actually institutionalizes mediocrity or even forces the adoption of bad practices.

There are several characteristics of the architecture of enterprise project planning in this next generation. The first is obvious: It's based on an enterprise project planning system, not a tool. The second characteristic is its hierarchical structure, particularly in the definitions of each level of this structure. The third is one that is all too frequently overlooked: a top-down rather than a bottom-up architecture. This leads to the next characteristic, the distinction between project planning and work management, which are frequently confused. Finally, the architecture for project planning needs to match the architecture for resource management, since the two are tightly integrated.

Enterprise System versus Project Planning Tool

Enterprise-wide project planning requires an enterprise-wide system for project planning. Although this appears to be obvious, it is often overlooked. It can't be done with a project planning tool where each project manager has her own project plan, which she keeps confined to her desktop. Enterprise project planning is a collaborative or shared architecture under which each member of the project team collaborates to develop a common plan and use this common plan to control all steps in the project at hand.

In an enterprise system, all project plans are stored on a common server and are accessible to everyone on the project team. Changes to the plan are made by those on the team with the appropriate authority to make those changes, and the system automatically reacts to those authorized changes in predefined ways. Other team members affected by changes are automatically made aware of them, so that they can adjust any dependent project steps and tasks accordingly. In other words, the project plan is a real-time distributed project control mechanism.

As we previously discussed, an enterprise project planning and control system within an integrated DCM system is very different from a planning tool, and this difference enables one to construct a management process architecture.

Hierarchical Structure

Project plans, especially for product development, are hierarchical. They consist of multiple levels, and each level expands on the one above to increasingly provide more detail. There is less detail at the highest level and progressively more detail at lower levels. Figure 10–1 illustrates the typical hierarchical structure for a product development project.

The top level defines the major phases of the project. Typically, there are three to six phases in a project, and project approval takes place at the completion of every phase, with a decision to continue the project, refocus it, or cancel it. The phase level is the strategic level for product development projects. It's the level at which senior management tracks the status of individual projects within the organization's portfolio of development pro-

> The phase level is the strategic level for product development projects. Project phases are consistently defined within a company, and all projects use the same phase definitions.

Project Planning Architecture

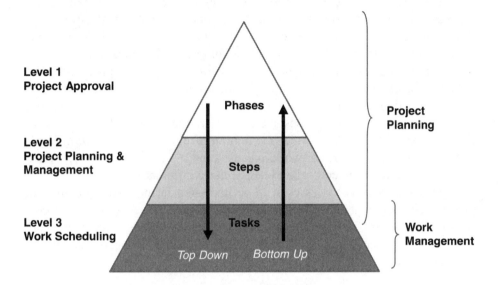

jects. In most cases, project phases are consistently defined within a company, and all projects use the same phase definitions.

Major steps are the next level within a project planning hierarchy (Level 2). Steps are used to schedule and manage the progress of development, and they provide the link between phases and tasks. This is the level on which the project manager and the project team focus to manage a project from end to end. A typical project may have 15 to 25 steps, spread across all project phases. Step definitions should be clearly defined, and then consistently applied to all projects.

Within this three-level architecture, steps play the most important role, since Level 2 is the primary level for project planning and management. This is true for several reasons:

1. Project planning for the total project is done primarily at the step level.

2. Planning standards for cycle times are usually set at the step level, since projects are configured with standard (and sometimes custom) steps. Steps in a project are like the standard components of a product.

3. Steps are the link between phases and tasks, and therefore between the strategic and detailed levels.

4. Most project management takes place at the step level; this is how progress is tracked and resources are scheduled.

The lowest level is the task level, although some project planning architectures may include an even lower level of detail that defines specific activities within each task. The task level defines the work that needs to be done, who is going to do it, and when it will be completed. Frequently, this level is further defined using a work breakdown structure that provides increasing detail. This work breakdown structure is sometimes referred to as high-level tasks, detailed tasks, and activities.

The top two levels—phases and steps—are the planning levels of a project management hierarchy, while the bottom level of tasks (or bottom levels of tasks and activities) are for work management. It's critical in a project planning architecture to understand and define these differences.

This hierarchy becomes even more important in the next generation of product development since its architecture establishes a common process across all projects. The hierarchy needs to be consistently defined, to work well for both project planning and project control, and to be clearly understood by everyone doing development.

Top Down versus Bottom Up

Project management architecture incorporates the fundamental assumption that projects are planned either from the top down or from the bottom up. The two are very different, and the choice of one or the other will define project planning practices and limitations.

In a *bottom-up architecture*, project activities are defined at the detailed task or activity level, the time required for each task or activity is estimated, and then the overall project schedule is estimated by "rolling-up" task and activity times, frequently using dependencies between tasks to determine when a task can start. The overall project plan then represents the summation of the time estimated to complete the required work. This "rolling-up" can be appropriate for relatively simple projects. In fact, some project-planning software applications are based on a bottom-up architecture, since this type of software tends to be designed for relatively simple projects.

In a *top-down architecture*, project planning starts at the highest level (the project approval level) with schedule-items expanded to the next level (project planning/management) in more detail. In this architecture, standards are used to set planned cycle times based on the characteristics of the project and the work expected to be accomplished within these standards, unless there are justified exceptions, such as a unique development step not done before.

Bottom-up project planning is the wrong architecture for complex product development projects, because it tends to encourage the accumulation of conservative cycle times, resulting in a longer time to market. Top-down project planning enables a company to set cycle-time standards for every step in order to achieve the shortest time to market.

The shift from a bottom-up approach to project planning to one that is top-down was one of the primary drivers of project cycle-time improvement in the TTM Generation. The new approach enabled companies to set cycle-time standards and achieve consistent time-based performance across all projects. The PACE process is an example of a top-down project planning process. PRTM has implemented PACE at hundreds of companies, achieving significant reductions in time to market (usually 40 to 60 percent, depending on the industry); top-down project planning was a very important element in obtaining these reductions.

> The shift from a bottom-up approach to project planning to one that is top-down was one of the primary drivers of product cycle time improvement in the TTM Generation.

Unfortunately, as was mentioned, some companies are making a critical error when they implement project management software that uses a bottom-up architecture, for they are unknowingly abandoning top-down project planning and its time-to-market benefits. The result may be that when they automate project planning, time to market will *increase*. These companies are confusing the work-breakdown structure with the planning structure.

Distinction between Project Planning and Work Management

Another important architectural concept for enterprise project planning is the distinction between the project-planning and work-management levels. As is illustrated in Figure 10–1, the upper part of the project hierarchy is intended for project planning. Phases are used for overall project approval, and steps are used as the central elements of project planning. In enterprise project management, these two levels must be universal across all projects, with the project plans shared across these two levels as part of an integrated DCM system. In some cases, high-level tasks may also be used for project planning, but planning in too much detail can be futile. Within the planning levels, it's important that the project plan be shared across all projects, and that it be accurate and complete.

Work management takes place at the lowest level, which includes detailed tasks and activities within tasks. At this level it's helpful but not essential that the schedule information be shared; it is not critical that it be accurate and complete. For some projects, this detailed schedule information may be useful for time management, but for others, it may not be; some individuals may even want to manage work at this level while others in the same project won't. Requiring consistency at the work management level across all projects is not productive, and in any event is hard to achieve.

For these reasons, it's necessary to clearly distinguish the planning level from the work management level. Typically, phases, steps, and optionally high-level tasks are included in the planning level, while detailed tasks are at the work management level. Enterprise project planning and control are done at the planning levels, not the work management levels.

Integration with Resource Management

Project planning must integrate with resource management, so it's essential that the two architectures be compatible. If integrated well, the two architectures map seamlessly to one another and fit within an overall architecture for this new generation of product development. Figure 10–2 shows how these architectures fit together. Project approvals are the focus of the highest level of both. Projects are approved phase by phase, and resources are assigned phase by phase in conjunction with the project-approval process. Portfolio and pipeline management takes place at this level, and this is where senior management focuses. To be successful in strategically managing projects, all projects consistently need to follow the same phase structure, and resources need to be assigned by phase.

Steps are the next level of the project planning framework, and they are also used for resource requirements planning in resource management. This level is the primary focus for project managers, who perform their project planning and resource management here, and the entire project team needs to apply steps in a consistent fashion. Resource-based project planning, discussed in Chapter 7, is enabled because of the relationship between these two frameworks at this level.

With resource-based project planning, project managers no longer assume the infinite availability of resources when planning projects. They can now plan the timing of project steps based on the availability of the resources needed to do the work at each step. Project planning is no longer based on starting every step as soon as possible regardless of resource availability. Steps can be timed earlier or later, depending on resource availability.

The lowest level of both frameworks focuses on work management. Detailed project tasks are defined in project planning, and detailed workload

F I G U R E   10–2

Integration of Project Planning and Resource Management Architectures

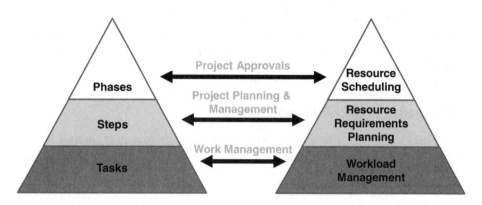

management is done in resource management. In both cases, the level of detail defined should only be what is helpful. It isn't necessary that all projects apply a consistent level of detail in the work breakdown, and it may not even be necessary to define the level of detail consistently within a given project.

THE ENTERPRISE PROJECT PLANNING AND CONTROL PROCESS

What is an enterprise project planning and control process? First, as a management process, it entails multiple development personnel (hundreds or even thousands of personnel, in some companies) working together in accordance with commonly understood practices. Second, it is a process enabled by a common system that everyone uses; in this case, the enterprise project planning and control system within a DCM system. Third, the process encompasses specifically defined project planning and control functions. Finally, it includes a clear definition of responsibilities and authorities for project planning and control, so that everyone understands how to work together.

The functions within this process include *collaborative project planning* by key developers on the project team, all working together to shape major project steps and tasks into a common project plan. Everyone working on the project team also helps to keep the project in control through shared responsibility for managing to the project plan. This *collaborative project control* is enabled by the interdependency of project schedule-items defined when the project was planned.

An enterprise project planning and control process enables a company to make *planning standards* an integral part of that process, and to reasonably expect that all teams will use the same project management best practices. It also enables consistent *integration of project plans* with other processes, particularly resource management, project budget management, and knowledge management. Finally, the enterprise project planning and control process provides a common platform for the *networked project team*.

Collaborative Project Planning

Collaborative project planning creates a common project plan shared by the entire project team, and it's the sharing of the project plan that transforms project planning from a technique into a process. However, allowing many people to jointly maintain a project plan, instead of leaving that job to the project manager's intuition, introduces the need to explicitly define the way in which project schedule-items interact with one another. The understanding of these interdependencies, and of how changes in one schedule item affect others, must be a common understanding.

> It's the sharing of the project plan that transforms project planning from a technique into a process.

There are several different methods for defining these interdependencies and relationships, depending on the enterprise project planning application software employed. These various methods involve defining the dependencies among schedule-items as well as how each item is expected to react to changes in another. The completion dates for schedule-items could change when the dates for a dependency change or they could remain firm in spite of a change to some other dependency. Planning interdependencies also involves defining how subordinate schedule-items (sometimes referred to as "children") affect the schedule for higher-level items (sometimes referred to as "parents").

Figure 10–3 illustrates some of these schedule-item interdependencies. Step A has four underlying tasks, with the second and third tasks (T2 and T3) dependent on T1, and T4 dependent on T3. These are all children of the parent Step A. The time needed to complete T3 is extended, and there are three ways that this could be interpreted by the project plan.

In alternative A, task T4 is "anchored," meaning that it is not automatically rescheduled based on a delay in completing the previous step. It starts at the same time as originally planned, and the lag time between T3 and T4 is eliminated. In alternative B, T4 is not "anchored," so it is automatically ex-

F I G U R E   10–3

Alternative Definitions of Schedule-Items

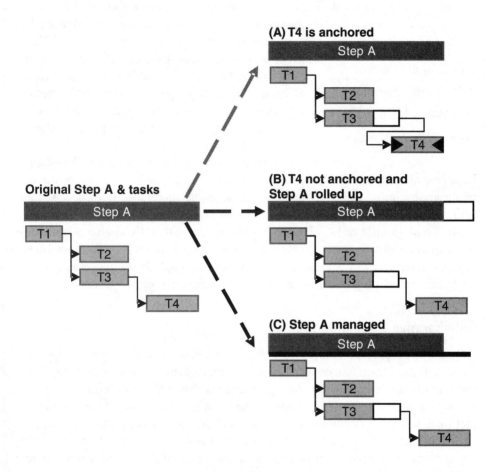

tended if T3's completion is late. The cycle time of Step A is rolled up automatically from its children, so its cycle time is automatically extended.

In alternative C, T4 is not "anchored," so it is automatically extended, similar to Alternative B. In this alternative, however, Step A is considered to be a "managed" parent item, instead of a rolled-up item, and therefore it is not automatically extended. Instead, the system indicates a discrepancy between the cycle time of the tasks and their parent, Step A, with an extended line.

The point here is not to explain project scheduling techniques, but to illustrate that there are scheduling alternatives that need to be explicitly defined in a common project plan. An enterprise project planning process establishes common guidelines for defining interdependencies. Everyone knows when it's appropriate to use each alternative and what is expected in terms of interdependencies. This process synchronizes project planning, and it helps individuals understand how their activities affect others in the project schedule, while helping them to understand management directives and other realities relating to the project schedule as well.

An enterprise project planning process establishes common guidelines for defining interdependencies.

Within a project team, a well-developed enterprise project plan enables project managers to define top-down objectives while giving individuals the flexibility to schedule more specific tasks. At the same time, it enables project managers to spot discrepancies between these objectives and specific tasks. The way schedule-items are defined also establishes the authorities among team members for changing schedules. Finally, a well-developed enterprise project plan allows project managers to use the project plan for collaborative project control as well as collaborative project planning.

Collaborative Project Control

Collaborative project control follows from collaborative project planning. Once a project team has created its project plan, then it can stay in control by collectively and individually managing against this plan. With enterprise project planning and control, everyone on the team has access to the common project plan, and each developer can track his or her progress against this plan.

Collaborative project control typically starts with the assignment of responsibility for individual project steps to members of the project team. These formal assignments are essential to the organizational dimension of the enterprise planning and control process. Each step has an "owner" who is responsible for managing that step. If the schedule for a step needs to be changed for some reason, the step owner makes the necessary changes and can see the impact on the project as a whole. The owner of the step can also see the impact on that step from other changes in the project. This hierarchical responsibility may be carried further down below the step level, to high-level tasks.

The step or task owners are also responsible for regularly reporting progress and completion percentage of their steps as part of the overall collaborative project-control process. The completion percentages of individual steps can then be combined with those from other steps to determine the overall completion percentage of the project as a whole.

When project control is collaborative, responsibility is distributed throughout the project to those who are closest to the work. At the same time, everyone is using the same system as part of a common process, and the system manages the interrelationships of the project schedule.

Integration with Project Planning Standards

Historically, many project management failures have stemmed from the lack of a formal, repeatable process for project planning and control. Project managers were left to their own devices to establish their own project planning and control process. Some project managers were naturally good at this task, some eventually learned to do it through trial and error, and others failed so badly in their first attempt that they never got a second chance. On top of this, projects were dependent on what was in the project manager's head. If the project manager didn't communicate her idea of the project plan, then others on the team were made inefficient or frustrated. If she left the project, the project plan usually fell into utter disarray. The lack of a clear and commonly understood project plan created an unnecessary risk of failure; a risk that can now be greatly reduced or even eliminated with an enterprise project planning and control process.

In a structured product development process like PACE, project steps provide the primary basis for cycle-time planning. Each step has a standard cycle time, and the configuration of the steps in a phase determines the cycle time for that phase. In many cases, companies establish cycle-time standards for each step, with variations based on the characteristics of that step. These standards have historically been published as project process guidelines that project managers are expected to follow, but this practice has some limitations. The standards are difficult to maintain, and not every project manager complies with using them.

With enterprise project management, these cycle-time standards can be easily maintained in step libraries and automatically applied in project planning. This is illustrated in Figure 10–4. The project manager selects the appropriate step from the step library and positions it in the project plan. In this example, the project team leader selects the step for version C of the business plan step. Similarly, she selects other steps from the step library. In a development chain management system, step libraries can provide a number of benefits beyond accurate, standard, cycle-time calculations. Keep that point in mind, since we'll be revisiting step libraries later on.

Selection of Steps from a Step Library

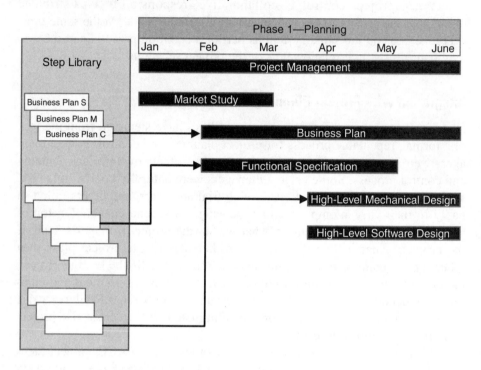

Integration of Project Plans

In an enterprise project planning and control system, the project plan serves a central role that extends beyond the project team itself. We already examined resource-based project planning in Chapters 5 and 6, so we won't revisit it here. Resource-based project planning is a major advancement from the previous generation of planning practices, which deliberately ignored the reality of limited development resources. Resource-based project planning requires an integrated enterprise project planning and control system; project plans on the project manager's desktop computer simply cannot be integrated with the necessary resource information. The project planning and control process expands project planning to include this integration.

An enterprise project planning and control system is also necessary to enable integrated project budgeting, since the budget is driven by the project steps and the resources assigned to them. We'll discuss this in Chapter 13.

The same is true for integration of project planning and control with knowledge management, which we'll discuss in Chapter 16. As we will see, the enterprise project plan provides the context for context-based knowledge management. We will also explore the opportunities for distributed program management, including the ability to plan and control a set of related projects as a program. These opportunities are also enabled by enterprise project planning and control.

Project Plan as the Basis for the Networked Team

In the next chapter we will look at the concept of the networked project team. The enterprise project planning and control process is essential to a networked project team, since this process is the most important coordinating mechanism. In fact, it's the central metaphor for the team, since project work follows project steps and tasks.

A subset of the project team self-organizes around each project step in order to complete it. In this self-organization lies one of the major benefits of the networked project team. The members of the step-specific team self-coordinate their activities, automatically sharing their individual work on the step as it progresses. The team members collaborate on project documents created in their step, raise issues that need to be resolved regarding the step, and coordinate action-items and detailed tasks among themselves. The step itself becomes a temporary virtual workspace within which the step-team members coordinate their efforts. In this sense, the enterprise project plan provides the intelligent platform for the networked project team.

ENTERPRISE PROJECT PLANNING AND CONTROL AT CRI

Anne Miller, the project manager of the Fast-Food Robot project at our hypothetical company, CRI, worked with her team to assign responsibilities for specific steps in Phase 1 (the planning phase) of the project. Richard Salisbury was responsible for the Market Study step, Naoto Takeyama for the High-Level Software Design step, and Lynda Stevens for the High-Level Mechanical Design step. Anne took responsibility for the three other steps herself. Figure 10–5 illustrates the Phase 1 step responsibilities, along with a few selected, high-level tasks.

Since this was the first time that most of her team had used the new enterprise project planning and control system, Anne explained what she expected. "Each of you has responsibility for managing your assigned step to the schedule, and you need to manage the high-level tasks within each step. You

F I G U R E   10–5

Fast-Food Robot Enterprise Project Plan

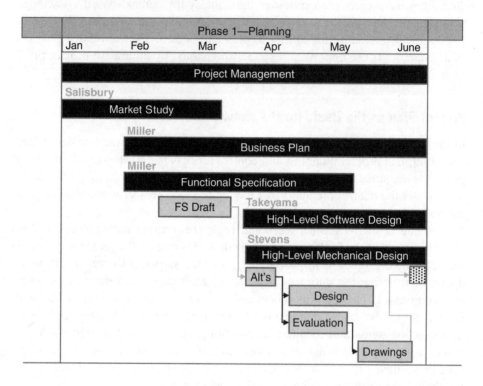

can define the high-level tasks as you choose, but I recommend that you use the standard high-level tasks from the step library for that step, and that you limit the number of high-level tasks for each step. It's easier to manage that way. You can also work together to define the interdependencies among steps or high-level tasks. I think it's better to make these dependencies with high-level tasks in order to provide a little more control."

She illustrated her point with the High-Level Mechanical Design step, which has four high-level tasks: identification of alternatives (Alt's); mechanical design (Design); technical, cost, and performance evaluation of alternatives (Evaluation); and preliminary mechanical drawings for the alternative selected (Drawings). Lynda Stevens, who was responsible for the step, also included a two-week buffer at the end of the step as a contingency. The Alt's step was dependent on the functional specification draft from the Functional Design step, which in turn launched the mechanical design work. The Design and Evaluation tasks followed the alternatives-identification task, and the Drawings task followed the Evaluation task.

As the planning phase progressed, the team ran into a difficult issue. Marketing could not agree on the type of mechanical delivery arm for the robot. This is the robot arm that picks up the food from the staging area and puts it on the tray in front of the customer. The team considered several alternatives. One of these, a delivery chute, was quickly discarded, even though this was what CRI's competitors were using. One of the team members summed up everyone's concerns. "A delivery chute would make it look like a vending machine, not a robot. We're trying to make our robot look like a person. We have the technology to do this right, and differentiate our robot from those of our competitors."

Two other alternatives for the delivery arm were viable, and there was a debate among the team, and indeed throughout the company, regarding them. The first alternative was called the "over-the-head dunk," where the robot's action would be similar to a basketball player reaching backward over his head. The robot's telescopic arm would extend directly over its head, select the proper food package, pick it up, and then retract and gently put the food on the tray. New food packaging would be required, but this would be necessary in any case. Some people at CRI thought that customers would find the "over-the-head dunk" motion entertaining, which would create immediate interest in the robot and help accelerate adoption. Others argued that this feature would violate the most important marketing requirement, which was to make the move in a somewhat human fashion.

The other alternative required the robot to move, instead of simply reaching out from a fixed position. It would take the order, turn around, move 2–5 feet, reach for the food, then turn around, move back to the counter, and place the food on the tray. While this motion might make the robot seen more animated, there was also the potential that it would make the device seem even less human. The cost of both alternatives still needed to be evaluated, and there was still some disagreement about the mechanical reliability of each alternative.

The general consensus was to do the high-level design for both alternatives and present them at the Phase 1 review, along with the possibility of CRI producing both versions of the robot. This decision added a lot of additional work to the high-level mechanical design, but Lynda Stevens came up with the appropriate changes so as to avoid delaying the project. As is illustrated in Figure 10–6, Lynda proposed starting the evaluation task three weeks earlier, based on the two preliminary alternatives, and extending the evaluation task another three weeks. The mechanical drawing task time would also need to be

F I G U R E 10–6

Fast-Food Robot Project Schedule Change

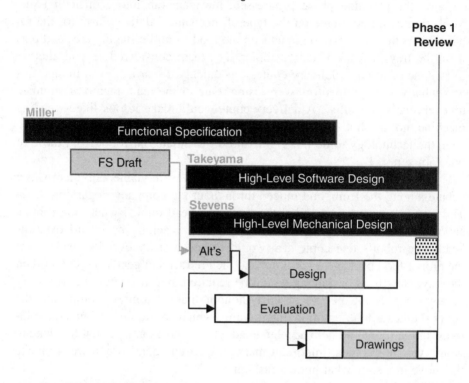

doubled, and she proposed starting it earlier and continuing it later, using the two-week buffer she had provided for. This solution seemed workable, and Anne approved the changes to the schedule. Lynda would still need to assess the impact of the additional costs, but thought that this could be done within the project's budget tolerances.

BENEFITS

Enterprise project planning and control have benefits that extend beyond individual project planning and control, including the following:

1. Collaborative project planning gets more out of the project team in terms of both their support for the plan and the plan's quality.

2. An enterprise project plan provides a common view of project schedules that is shared by the entire project team.

3. Project planning standards are more easily implemented through enterprise project planning than through published guidelines, thereby improving time to market.

4. All updates to the project schedule are reflected immediately in the project plan and communicated to everyone on the project team, reducing miscommunication and misunderstanding.

5. Distributed responsibility for project steps and high-level tasks increases accountability and predictability, and therefore improves project execution.

6. More predictability, increased consistency, and lower risk result from a common enterprise planning and control process.

7. An enterprise project plan also provides the foundation for other capabilities, such as resource-based project planning, the networked project team, enhanced phase reviews, and distributed program management. We'll examine these capabilities in the following chapters.

SUMMARY

We began this chapter by recognizing the important role of structured development planning in the cycle-time and plan-reliability improvements achieved by the Time-to-Market Generation of product development management. We then looked at the use of project planning tools by project managers, the limitations of those tools, and how they are being improved by the next generation—the R&D Productivity Generation—of product development.

The central focus of this chapter was on the transition of project planning and control from a technique into a management process: the enterprise project planning and control process. We discussed its architecture, including an enterprise system, hierarchical structure, top-down versus bottom-up planning, and the distinction between project planning and work management. We also discussed the major functions of this process, including collaborative project planning and control and the process's integration with other development chain management systems and processes. Finally, I tried to bring enterprise project planning and control to life with an example of its use at our hypothetical company, Commercial Robotics Inc.

To reiterate, the enterprise project planning and control process represents more than an overt change to where project-level information resides within development, and how that information is disbursed to members of the project team. Perhaps even more important are its constructive effects on the attitudes and behaviors of project team members. By making project planning and control a collaborative affair, team members are driven by a sense of ownership to make the plan work at the crucial level of its individual project steps and high-level tasks. This sharing—this invitation to participate in the life of the project instead of just following directives—takes project leadership to a higher plane of effectiveness.

11

CHAPTER

Networked Project Teams

Project team organization models have evolved over the last 50 years, and their evolution has been an important part of each new product development generation. In the late 1980s, the cross-functional core team model proved to be so effective that it replaced the functional team model as the standard. Now, a new project-team model has emerged that surpasses the cross-functional core team in terms of performance. I refer to this as the *networked project team.* The members of a networked project team are able to communicate, coordinate, and collaborate with one another across geographies and time zones. The earlier constraints on collaborative development formerly imposed by distance and time are eliminated, and the inefficiencies of using email are replaced by more effective communications mechanisms.

Figure 11–1 illustrates the evolution of three major project-team organizational models over time. The functional team was the primary model from the 1950s until the late 1980s. It accorded with the strong command-and-control hierarchical structure of companies during those decades, and was the first formal project team organization model. Each function completed its particular work on the product under development, and then "threw it over the wall" to the next development function. This rigid segmentation and sequencing of the development process, so reflective of the assembly-line mentality of the time, all too often resulted in poor products and long product development cycles.

F I G U R E   11-1

Evolution of Project Teams

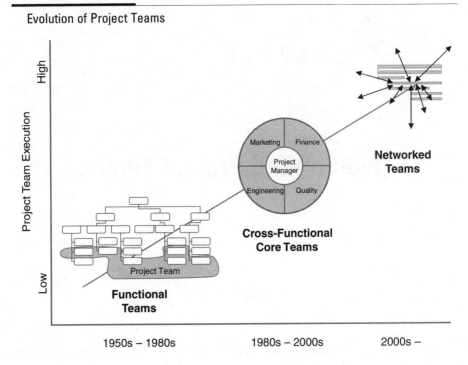

The functional team structure had many vertical layers and generally lacked horizontal communications, slowing coordination and time to market, because key decisions first had to flow up the vertical organization for approval and then back down again. The primary defect in the functional project model was in the structure itself. It erected unnatural barriers among functional groups in what was inherently a cross-functional process.

In the 1960s and 1970s, several variations of this functional model emerged as companies tried to correct these deficiencies. Matrix teams, program coordinators, and dedicated project management offices were examples of such variations. In the end, however, these adjustments to the model could not overcome its fundamental flaws.

CROSS-FUNCTIONAL CORE TEAM MANAGEMENT

The next major advance in project management organization was introduced in the late 1980s, with the Time-to-Market Generation. The cross-functional core team organizational model and related core-team management practices enabled project teams to better synchronize their execution of project steps and tasks by *organizing* the teams more effectively. The simplified organizational structure of the core team enabled dramatic improvements in project communication, coordination, and decision making.

The original cross-functional core team model was optimized for time-to-market reduction, and it was a key factor in reducing time to market by 40 to 60 percent at many companies. Core-team best practices included colocating the team members and dedicating them to a single project for the project's duration. While these practices enabled constant close communications and rapid decision making, they also imposed some undesirable restrictions. There was virtually no flexibility regarding where and when the team members worked, or how their time and talents were used by the corporate development organization as a whole. By the mid-1990s, many companies adopted email as a solution to these restrictions. But in my opinion, the uncontrolled and unstructured nature of email communications has undermined much of the original core team effectiveness and actually decreased project productivity. I'll explain my view in some detail, but first let's review the basic concepts of the cross-functional core team, since we'll be building on them.

> The original cross-functional core team model was a key factor in reducing TTM by 40 to 60 percent at many companies.

Traditional Cross-Functional Core Team Model

The high degree of uncertainty and variability in product development underscores the importance of good communication. Effective project communication needs to take place both vertically and horizontally, and the cross-functional core team organization replaced a chain-of-command functional team structure that was both cumbersome and error-prone. In the cross-functional core team model, a core team of five to eight managers acts like the hub in a hub-and-spoke design and is directly responsible for the success of the project. As illustrated in Figure 11–2, managers communicate very closely with one another and manage all communication with others in the extended project team for whom they are responsible. In some cases the core team member directly manages a group of developers assigned to the project, and in others the core team member has responsibility to coordinate project work with others outside of the core team, but everyone working on the project communicates through a core team member.

F I G U R E 11–2

Core Team Structure and Communications

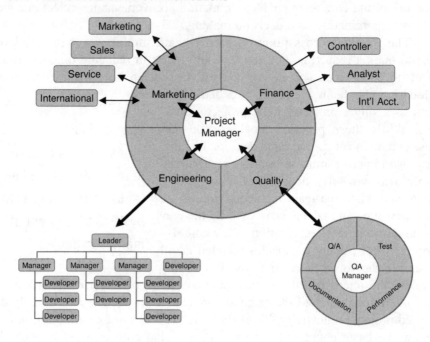

This organizational solution solved many communications problems. Because the core team worked directly together on a daily basis, they were able to communicate rapidly and reliably. No information was lost among the team because of organizational filtering or distortion. The hub-and-spoke arrangement makes for highly reliable team communications. Figure 11–2 represents a typical core team of five, four representing the major disciplines and project requirements, and one project manager. Simple and reliable communications are especially important in a large project, because the number of possible communication paths among those on a project team increases geometrically with the number of people involved in the project. With just 30 people on a project team, there are more than 1,000 possible communication paths among them. The core team organization defines two structured communication paths: communication of almost everything within the core team, and communication to the extended team through the designated core team member.

Developing a new product requires the completion of thousands or even hundreds of thousands of activities, and many of these activities are interdependent. Ineffective coordination of these interdependent activities is a major reason for project delays and wasted time. It's often the case that one person's activities can't begin until someone else's are finished. Sometimes an activity in a sequence of activities is delayed because the person assigned to perform it isn't told that the previous activity in the sequence has been completed.

The cross-functional core team organization enables the effective coordination of project team activities, with minimal wasted effort. The core team is directly empowered, and made directly responsible for the success of the project and all its steps and activities. The team members typically meet regularly to coordinate project activities among themselves, following which each member coordinates the activities of his or her extended team members. The hub-and-spoke organization of the core team thus works as well as a coordinative mechanism as it does as a mechanism for communication.

As mentioned previously, cross-functional core teams originally performed at their best when all their members were colocated and were assigned to one project to the exclusion of all others. These twin concentrations—one place/one project—meant that the core team could meet almost immediately whenever necessary, quickly resolving issues and immediately coordinating activities.

Most companies were able to cut their time to market in half during the 1990s, and the advent of the cross-functional core team was a major reason for that achievement. Nevertheless, the rigidity of the one place/one project core team arrangement obviously had some frustrating drawbacks, particularly with the rise of the Internet.

Distributed Email Communications in Core Teams

By the mid-1990s, new communications technologies such as email and voicemail made physical colocation seem less critical, and many core teams began to use these technologies to gain more freedom of movement and scheduling. While email and voicemail did provide some welcome flexibility, they also created new problems, which continue to drag down project team productivity and in some cases have made team communications inefficient and unreliable all over again. As communications came to be distributed among all project team members, there was little or no structure to the team's communications. The control of critical project content diminished, and a significant administrative burden was added to the work of all team members. Some teams substituted email for verbal communications, and much was lost in the translation.

Email certainly simplifies communications from the sender's perspective. Emails are easily sent to many people on the project team, or even the entire team, as well as those outside of the project. But the very ease of sending out emails has opened up the communications path issue all over again. For example, to whom should an e-mail with a technical specification be sent? Too often, the easiest answer is, "Just send it to everyone, or almost everyone, just to cover all the bases." But now, everyone receiving the email needs to read it, process it, and file the attachments on their own computers in some organized way. Even if the email isn't relevant to them, they still need to determine what to do with it once they receive it. Sometimes, responding to emails that are not important or not a priority when received distracts developers from priority tasks. In a modern company, one could easily spend half one's time doing one's work, and the other half responding to company emails. The phrase "gratuitous communication" comes to mind to describe this.

> In a modern company, one could easily spend half one's time doing one's work, and the other half responding to company emails.

Some informal estimates indicate that it is not unusual for each member of a large project team to receive 30 to 50, or even more, project emails daily, and to spend one to two hours a day just processing and organizing this email traffic. This is repetitive, wasted time: Everyone on the team ends up sorting through and organizing the same emails and attachments, wasting hundreds of man-hours every month. If, for example, there are 30 developers on a project team and each developer spends just an hour per day processing project emails, that's 600 wasted hours every month. This equates to approximately 3.5 wasted developer years over a 12-month period—more than 10 percent of the project's assigned resources and a waste of as much as $600,000 for the year on the project.

Conversely, it will not come as news to experienced R&D managers that some development personnel simply ignore their incoming emails, letting the messages and documents pile up. While these renegades do not waste the time of those who faithfully attend to the daily torrent of project emails, their failure to process important communications can play havoc with project planning and control as well.

Figure 11–3 illustrates the perils of distributed communications. Each person on the 32-member team pictured in the diagram can send an email to any or all of the 31 others, expecting them to process it, understand it, and either file it or act on it. Each line in the figure represents the communication of one person to another. While it looks confusing, it's actually understated; otherwise, the entire

F I G U R E   11–3

Core Team with Distributed Communications

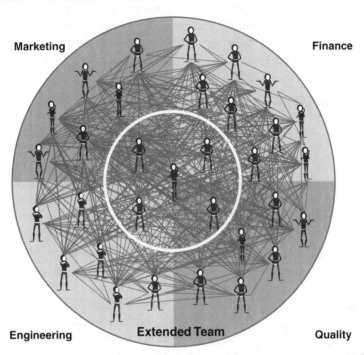

figure would be black with communication lines. With so much potential for unstructured and uncoordinated communications, it is inevitable that some people who don't need to be involved in a specific matter are brought into the communication loop anyway, while others who need to be involved are left out.

An additional drawback is loss of content control. Various project team documents are distributed throughout the team. Nobody is quite sure where the "official" copies of documents are located, or who has the latest complete set of documents. Eventually, companies that allow this situation to develop will find that they have unintentionally "lost" some of their most critical project documents and intellectual capital.

Finally, the original coordinating mechanism of the core team will also deteriorate when the communications function is distributed throughout the entire team. At best, more and more coordination ends up being done outside of the core team, making it easy for the team to lose some degree of control. At worst, the core team simply dissolves and loses control of the project.

Project network systems now enable project teams to transcend the limitations of time and distance while providing the necessary coordination mechanisms, content control, and structured communications that are missing from email-based distributed communications. Project network systems improve project team execution by *networking* team members, enabling a new level of improvement in project execution.

THE NETWORKED PROJECT TEAM

The networked project team is enabled by a project network system based on the enterprise project planning and control system we discussed in Chapter 10. This system provides a virtual project colocation environment. Visualize this virtual environment as a giant conference room that holds the entire project team. On one wall is a big project schedule, where team members attach the latest version of the work they are doing on specific project steps and retrieve anything of interest attached to the schedule by other team members. On another wall is a large calendar, with every project schedule-item of interest to the team. Team members also attach documents to the calendar for every meeting and event, so that anyone can go and find documents and information they need. Another wall is used by the team to post messages to one another. Team members can leave messages for others by name, or (more usefully) they can leave messages by topic or schedule-item. The team members working on a project step will regularly review any posted messages related to that step, and anyone interested in the latest version of project documents for that step will know when the documents have been revised. The "situation room" is never closed, and team members stop by regularly throughout the day. Now, imagine that there is a separate situation room for every project, and that developers, in the course of the workday, routinely visit the various rooms connected with the different projects in which they are involved. This is the environment provided by a project network system—only it is a virtual one, instead of a physical one.

> A networked team using a project network system operates very differently from a core team using distributed email communications. Unlike the first, which communicates through a physical hub-and-spoke design, the second communicates dynamically.

To elaborate: Today, the key to successful product development teams lies not in how the teams are organized, but in how they are *networked*. The networked project team communicates much more coherently, enabling project team members to self-organize around specific-project work elements. The networked project team more effectively coordinates its activities by using a virtual project team environment based on clear mechanisms for coordinating project communications and activities. It efficiently manages project content, freeing developers from this redundant administrative task.

A networked team using a project network system operates very differently from a core team using distributed email communications. These distinctions are seen in the differences in the way project team members communicate, in the project-coordinating mechanisms provided by the project network system, and in the way project content is managed.

PROJECT COMMUNICATION IN NETWORKED PROJECT TEAMS

Unlike the cross-functional core team, which communicates through a physical hub-and-spoke design, network teams communicate dynamically, but yet in a structured way. The problems of what to communicate, whom to communicate it to, and when to communicate are resolved, for the most part. Communications can take place at a time that is convenient for someone posting information, and at a different time that is convenient for someone else interested in that information. Project team members working together no longer need to be colocated, since they now conduct their communications within the virtual project environment.

But unlike distributed project communications using email, communications in the networked project team are more structured. The project network system establishes an orderliness to communications that is missing from distributed email communications. Networked team members who are involved in a project step or event communicate with one other without distracting those who are not involved with unnecessary communications.

The structure of communication in a network comes from the communication switching provided by a centralized project network hub, the project network system, instead of from the router-like model used for email-based communications. All project team members communicate through the project network hub by adding information to the project network system as they do their project work, and by obtaining any information from the project network system as they need it.

These structured communications enable one of the important characteristics of the networked project team: Project team members can self-organize around any piece of work. They self-organize through their interest in communicating on a particular project step or other similar coordinating mechanism.

Those working on a step are obviously interested, but others can also self-select to communicate regarding this step. In a traditional cross-functional core team, the core team either coordinates all communications or formally designates a group within the team to do the coordinating.

For example, CRI's Fast-Food Robot project team really appreciated the more efficient communications that came from being a networked project team. In Phase 2 development, the team consisted of 45 developers, representing a wide range of skills. Ten of the developers worked on designing the robot's voice-processing module, a step within the development phase, and they used the project network system as a mechanism to self-organize their work around this step. They maintained all the detailed design documents for the voice-processing module on the project network system by linking the documents to the voice-processing design step. The voice-processing module team also found it useful to attach to that step all the technical documentation for the signal-processing components, the microprocessor, and the main software kernel. This way, any one of them, or anyone on the Fast-Food Robot project team, for that matter, could easily locate current technical information whenever they needed it. The electrical designs were done using CAD and CAE software, with all of the drawings linked to the voice-processing design step.

One of the main tasks of this step was to collect voice samples from actual customers. Marketing had an analyst in the field recording samples from a number of McDonald's restaurants, and these audio recordings were used to test the early versions of the voice-processing module. The results of these tests, as well as the audio recordings themselves, were linked to the voice-processing design step.

The 10 developers assigned to this step worked together naturally. There was no need to assign a formal subteam or hold frequent meetings. They self-organized around the voice-processing module design step in the virtual environment we discussed. They performed their work in this virtual environment, communicated their ideas and concerns regarding this common step, and associated every piece of information regarding the voice-processing module around this step. Yet, at the same time, this step was also part of the total Fast-Food Robot project.

The 10 developers working on the voice-processing module were based in six different countries and worked in time zones separated by up to 12 hours, yet this never seemed to matter very much. They were in constant communication and collaborated closely. Half of these were also involved in other steps of the project, and were part of other self-organized teams within the Fast-Food Robot project team.

At the same time, everyone else on the larger project team who needed to stay abreast of the voice-processing module work could easily do so by viewing the communications and documents of this step of the project. Anne Miller, the project manager, would periodically check up on this step in the virtual team environment to keep abreast of what was going on. If there were any results that concerned her, she used a note or action item to communicate with the voice-processing design group.

Anne found many advantages of the networked project team over the previous core team she managed. Communications were a lot more efficient. She now spent very little time doing email administration, freeing her time to better understand the product her team was developing and spend more time face-to-face with team members. Anne also held fewer core team meetings, although, since this was her first experience with a networked project team, she reduced the number of meetings gradually as she gained confidence in the new model. She also felt that the team worked much more efficiently and got a lot more work done because communications were smooth and immediate.

> The networked project team reduces the need for dedicated core team members; it also makes it feasible for core team members to simultaneously participate in multiple project teams.

Networked teams reduce the communication burden on the core team, which was previously the primary communications-control mechanism for a project. There is much less need for core team members to meet regularly in person or by conference call. The networked project team also reduces the need for dedicated core team members. Because they can actively manage a project on their own schedule instead of a common core team schedule, it is more feasible for core team members to simultaneously participate in multiple project teams. Since this represents a behavioral change for core team members, it's important that the networked project team process be properly structured and that the teams be appropriately trained in order to get the best results. Simply implementing project networking software is not sufficient.

PROJECT COORDINATION IN NETWORKED TEAMS

In the networked project team, project activities are coordinated through the virtual project team environment using various coordinating mechanisms, such as scheduling elements of the enterprise project plan, project calendars, action-items, and so on. These coordinating mechanisms provide the intelligence

Coordination Mechanisms in Networked Project Teams

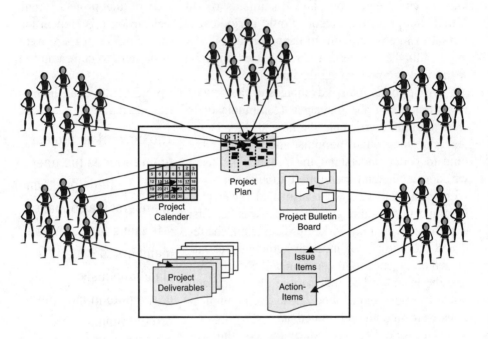

that enables team members to find what they need when they need it, to find out what they need to do, to determine if something they're waiting for has been completed, etc. There are various types of coordinating mechanisms provided by a project network system, and they may be used in different combinations depending on the preferences and practices of the networked project team. Some of these mechanisms are illustrated in Figure 11–4.

Project Schedule-Items

A project schedule is usually the central metaphor for the networked project team, and project work generally flows around major schedule-items, such as project steps or high-level tasks. This is another reason why the enterprise project planning and control process we previously discussed is so critical. A subset of the project team works on each step, and self-organizes around that step, providing one of the major benefits of the networked project team. While they are working together on that step they coordinate all their activities on it, automatically sharing their individual work on the step as it progresses. They collaborate on project documents created in that step, raise issues that need to be resolved regarding the step, and coordinate action-items and detailed tasks among themselves. For those working on the step, the step itself becomes their temporary virtual workspace within which they coordinate their work.

On the Fast-Food Robot project, there were seven people involved in the functional specification step, and they self-organized around this step. They coordinated their work by defining high-level tasks for evaluating customer feedback, evaluating competitive products, analyzing market expectations, developing a preliminary functional design, and then completing the functional specification. The team collaborated on writing the functional specification by individually adding to each version as it evolved. They identified functional specification issues as they went along, and coordinated the resolution of those issues. They also routinely initiated action-items for one another in order to coordinate the step's completion.

When a networked project team uses a schedule-item such as a project step to coordinate the work of team members, the item becomes more than just a schedule-item. It becomes the basis for the team to self-organize its work.

Project Calendar-Items

A project calendar can also serve as an effective coordinating mechanism for the networked project team. Everyone on a project team can easily think in terms of a calendar metaphor, so the team can be effectively coordinated around project dates, both past and future. Some networked teams may even use coordination around project dates as the primary coordinative metaphor, while others will use it as a supplement to schedule-item coordination.

The Fast-Food Robot project team scheduled its second product-design review for May 5th and placed this date on the project calendar. While those directly involved were scheduled to participate, placing this item on the calendar communicated to all other team members that this review was scheduled so that they could plan and act accordingly. When the meeting agenda was drawn up, it was attached to the scheduled date on the project calendar, as was the second revision of the product design specification.

When the participants were preparing for the meeting, they simply went to the online project calendar and printed out or viewed the agenda, along with the new design specification. There was also an online link to an article on a new competitive product that the marketing manager had brought to everyone's attention because he thought it might be germane to the design review discussion. Anyone who needed to check back to the previous version of the Fast-Rood Robot's product design specification could simply go back to the previous design review meeting on the calendar and download the specifications used at that meeting.

The manufacturing engineer wasn't planning on attending the design review, but checked to make sure that the latest manufacturing requirements were included in the revised specification. When he found that they weren't, he contacted the engineer running the design review, provided the relevant edits to the specification, attached these to the meeting materials, and also attached the new manufacturing guidelines as a reference in case they were needed for the review.

Using the project calendar as a control mechanism may be the easiest way for a project team to make the transition to a networked project team, since it requires less initial setup than using project schedule-items. The project calendar may also prove to be a sufficient coordinating mechanism for smaller projects.

Project Deliverables

Project deliverables provide yet another mechanism for project coordination. Some teams may prefer to focus on project deliverables as the primary mechanism for coordination, while others may use it as a supplement to project schedule-items. If project deliverables are used as the primary mechanism, then members of the networked project team self-organize around them.

The Fast-Food Robot project team used project deliverables to coordinate its pilot customer briefings, which were delivered to its primary pilot customers, McDonald's and Burger King. These briefings were given approximately every quarter, but were not major schedule-items or formal meetings. Anne Miller, the project manager, took responsibility for the briefing documents, but everyone on the team contributed to them, making the job much easier for her.

If used secondarily to project schedule-items, then it's useful to define key deliverables within a project step in order to focus coordination around deliverables as intermediate units of work and coordination within a step. As we saw in the above example, a deliverable can also be used as a coordinating mechanism, even when it doesn't fit within a schedule-item or calendar.

Project Action-Items

A typical project team needs to coordinate hundreds or even thousands of action-items throughout the project. An action-item is a task that needs to be done by someone on the team whose work is coordinated with the work of an-

other team member. It is typically something that one team memb
do so that another member can do what he or she needs to do. An ac
might be something as simple as a request to answer a question. Actio.
also arise from team meetings as agreed "to do's." Every action- item is assigned to someone on the team, and each item has an expected date of completion. Action-item management is an important
supplementary coordinating mechanism, especially when action-item administration is fully
automated.

> The project network system manages action-item administration. It keeps track of open action-items by person, initiates reminders when completion is overdue, and organizes action-items by category.

The project network system manages action-
item administration. It keeps track of open action-
items by person, initiates reminders when completion is overdue, and organizes action-items by
category. Anne Miller used action-items extensively to coordinate work details among team
members. She insisted that action-items be collected following every meeting, and she regularly added dozens of action-items each week for
various team members. She promoted a culture
in which everyone on the team could initiate an
action-item for another team member when he or
she needed something, but she also established guidelines for setting
item-completion dates that enabled sufficient flexibility.

Every Monday morning, Anne reviewed open action-items that were
overdue, as well as action-items by type and project step. This enabled her to
stay on top of emerging problems and gave her additional insight into project
work. Her team quickly developed the practice of clearing up as many action-items as they could prior to her Monday morning review.

Issue-Resolution Items

Issue-resolution items are a variation of action-items, and are created when a
collaborative effort, entailing multiple activities, is required to resolve an issue. Completing an issue-resolution item is more complex than completing an
action-item, so the progress toward completion needs to be tracked.

The Fast-Food Robot project team, for example, needed to determine the
performance requirement for the voice-processing module. They identified this
as a technical issue and placed it in the issue-resolution process for the project
network system to manage. Anne Miller, the project manager, assigned the software engineer to coordinate resolution, set the date by which the issue needed to

.e resolved, and defined activities to resolve it. Input on this issue was accumulated on the project network system. At any time, anyone on the project team could monitor the progress on resolving this issue, review working documents used to collect customer input, and review results of performance tests, as well as other information of this nature. In all, more than 20 members of the project team from all over the world, including some specialists who were only involved with the voice-processing aspect of the project, self-organized to collaborate on resolving the issue. Few formal meetings and in-person work sessions were needed, since the project network system made for efficient collaboration.

Issue-resolution in the networked project team works well because the members of the team and other specialists self-organize around resolving the issue. Their participation is free-form, but the project network system provides the basis for organization, structure, and coordination.

Project Bulletin Boards

A project bulletin board or similar posting function works best for informal communications and coordination that are not associated with a project schedule-item, calendar-item, action-item, or deliverable. General team notices are typically posted on a virtual bulletin board maintained on the project network system.

The bulletin board for the Fast-Food Robot project contained the following recent notices:

- A notice of an article from an industry journal, with a link to that article
- An introduction of a new project team member, with his profile
- A photograph of the financial analyst's new baby
- Notification that the design review meeting location had changed
- A link to a copy of the company president's speech at a financial conference
- A notice of a critical software bug found in the latest product test release, with a request for team input regarding the bug
- A notice concerning a recent meeting with a pilot customer, along with the minutes of that meeting

Members of the networked project team access the virtual team bulletin board in various ways. They may scan all recent notices periodically to see what's happening of interest, or they may examine notices by specific categories of interest. A team bulletin board is a much more efficient way of organizing general notices than sending emails for everyone to process.

CONTENT MANAGEMENT IN NETWORKED PROJECT TEAMS

The work of product development teams is frequently communicated through various forms of content, such as text, presentations, spreadsheets, drawings, pictures, and notes. The networked project team manages all this content through collaborative content management. Project content can generally be referred to as documents or files in one form or another, including project requirements documents, product specifications, project plans, product drawings, bills of material, marketing plans, pricing analyses, competitive analyses, progress reports, phase-review presentations, financial estimates, and so forth. Some of these files are text documents and others are charts and graphs. Documents may be made accessible to all team members or only to selected members, and some team members may have the application software to manipulate and edit these documents or files, while others may be restricted to a read-only format.

In the networked project team, all project documents are stored on the network and are accessible through the project network system, which manages all project documents. This is a major change from the email-based project team model, where project documents were physically distributed across the desktop and laptop computers of the project team members.

Project Document Management

Project documents are those created by and managed by the project team. On a moderate-sized project there could be hundreds of documents, and on a large project there could be thousands or tens of thousands. Networked project document management has several characteristics.

First, all project documents are filed on the project network instead of on the individual team members' computers. This provides control over all project documents and eliminates the need for every team member to process and file a copy of every relevant project document. Instead of a team member trying to decide whom to copy on a specific document, he simply files it on the network so that anyone who needs it will have access to it. Team members can immediately access any document without having to search through their computers to figure out where they filed it, and they don't need to request the current copy from another team member, who in turn needs to interrupt his work to find and send the document.

Second, documents are filed on a project network system very differently than they're filed in a file server. A file server simply organizes documents in folders, and anyone trying to find a particular document needs to sort through the folders and the list of documents by name, hoping to find what he wants. In a project network system, documents are generally accessible based on the coordinating mechanisms previously discussed. This can entail attaching project documents to a project step or task so that everyone working on that step or task can easily see the documents associated with it. Or it can be done by attaching documents to the project events listed on a calendar. Presentation materials and minutes can be attached to a meeting noted on a calendar. Design documents can be attached to a design review session, while presentation materials, project plans, and progress reports might be attached to a project review.

> The project network system automatically manages all documents and their backup and archiving, thus preserving the intellectual capital of the company's R&D, otherwise lost when the project content is distributed on the desktop computers of team members.

Third, revision control is maintained by the project network system. Every team member sees the current version of all documents, and has access to all previous relevant versions. This avoids a situation that arises all too frequently, where one team member is working from an outdated version of a critical document such as a product specification or drawing.

Fourth, the project network system also provides the context for accessing all product data-management documents and files. These are linked directly to the project steps to which they're related. In some cases, the link may be to a product data-management subsystem. Team members with the appropriate applications software to perform tasks such as modifying drawings can directly update these files and documents.

Finally, the project network system automatically manages document backup and archiving. This preserves the intellectual capital of the company's R&D, which is otherwise lost when the project ends, because all the project content is distributed on the desktop computers of the team members. Sophisticated project network systems can even archive these documents directly into the appropriate knowledge management repository.

Collaborative Document Preparation

In the networked project team, many project documents are prepared collaboratively. The talents of many on the team contribute to the initial drafting, revision, and completion of a document. Edits and changes need to be tracked by team members, and all these edits and changes need to be consolidated into the next version of a document. The project network system manages this version control automatically, and supports collaborative document preparation by managing version control of project documents.

Document management capabilities, combined with the appropriate procedural guidelines, can enable efficient and reliable collaboration, even simultaneously with multiple team members making changes. Automatic document management eliminates the need for teams to go through edits to a document together at meetings. It avoids the problem of losing control over edits when they're made to different versions of a document, and greatly simplifies the task of consolidating edits.

The Fast-Food Robot project team used collaborative document preparation extensively, developing document management practices and conventions that eventually became routine. They found that this eliminated the need for the document revision meetings they used to suffer through, and increased team productivity. Anne Miller found it particularly helpful and comforting to be able to easily find the latest version of every project document when she needed it.

NETWORKS OF NETWORKED TEAMS

The networked project team can also seamlessly configure into a network of networked teams. A major new product-platform project, for example, could consist of multiple networked project teams: the platform team, the team developing a new technology critical to the platform, and product development teams for the initial products that will incorporate the new platform. This ability to interweave networked teams is particularly important for complex projects. A project could be structured as a step in a broader project or as a separate project. In either case, developers working within this network of networked teams can coordinate across team boundaries, self-organize to address project issues or common work requirements, and dynamically share content and deliverables. We'll review this in more detail later when we discuss distributed program management.

Since the Fast-Food Robot project was part of a broader development program at CRI, it was part of a network of such teams. The language-processing module was directly related to a new technology project that was being implemented simultaneously on two other robots. For work on that module, the Fast-Food Robot team collaborated with the R&D team developing the technology, as well as with the other robot development teams. Through this collaborative networking they learned that the project team doing the Retail

Robot had very good success in using a scanning verification feature to handle voice-recognition problems, so they considered something similar for the Fast-Food Robot. They also used the network of networked project teams to self-organize around the resolution of a performance requirement issue pertaining to the voice-processing module, since the company wanted to use a common microprocessor.

By dynamically networking project teams wherever helpful, a company can much more easily conduct—and coordinate—parallel product development efforts. Possibilities in terms of networks of networked projects include the following, for example:

♦ Multiple project teams advancing common technologies or using the same underlying product platform

♦ Developing a technology in conjunction with the product that will incorporate the technology

♦ Developing a manufacturing or process technology in parallel with the product to be manufactured using the technology

♦ Developing new information systems in support of a new product being developed

♦ Supporting projects by suppliers or codevelopers aimed at providing modules or components for a particular product

REQUIREMENTS

The networked project team is a very compelling concept and will most certainly be the model for project teams in the next generation of product development. However, successful implementation of networked teams requires several ingredients.

First, it requires the appropriate project network application software. We've already seen that email-based applications are certainly not the answer. In fact, email-based distributed communications create as many, if not more, problems than they solve. Simply using a network file server that stores documents in folders on a common network is in no way adequate. Project network software needs to include the appropriate coordinating mechanisms for communications and content management. This requirement can only be met by the enterprise project planning and control system we discussed in the last chapter.

Second, the networked project teams need to be governed by policies, procedures, and practices that define such matters as how the teams will use the coordinating mechanisms, the procedures for document management, etc.

These policies and procedures need to be developed to fit a specific company's development profile, culture, and management, and they need to be systematically implemented. Frequently, this development and implementation are part and parcel of implementing the enterprise project planning and control process previously discussed.

Finally, implementing networked project teams requires a cultural change. Developers are used to the way a cross-functional team works, but they now need to learn how the networked project team works. They need to learn the principles and expectations of self-organizing work teams. Once they are comfortable with the change, they will find it much more rewarding and will be less frustrated by wasted time.

BENEFITS

The networked project team, self-coordinating by virtue of its coordinating mechanisms, has many performance advantages over the cross-functional core team using distributed email communications. Collectively, these advantages can significantly increase the productivity of product development teams, improve execution, and reduce errors. These advantages are worth reviewing once more:

1. The coordinating mechanisms of the networked project team greatly reduce the amount of coordination that needs to be done by core team members. This can free up a lot of time for core team members, enabling them to spend more time developing products.

2. The project core team no longer needs to be physically colocated. Members can work continents apart and still maintain the appropriate control over communications.

3. Project core team members no longer need to be dedicated to a single project. They can work simultaneously on multiple core teams. They no longer need to conform their work schedules to those of their fellow core team members, since every project has its own "virtual situation room," open 24 hours a day.

4. By centralizing content management, the networked team eliminates the administrative time wasted by project team members in processing, organizing, and filing the same project documents. On some projects, simply processing emails and organizing and filing the contents for later review takes as much as two hours per project team member each day. Eliminating this clerical labor frees up a lot of time for more productive development work.

5. On the networked project team, developers self-organize around work elements such as project steps, issues, and deliverables. Self-organization is highly productive because it is dynamic, extremely efficient, and reliable. The resulting productivity savings and quality improvements can be substantial.

6. Communications among members of a networked project team are fluid, but structured. Distributed communications using email are fluid enough, but they're uncontrolled. Project information in this previous generation was emailed to people to whom it applied only peripherally, or not at all, while people who needed the information might never have received it or never read it. There is less risk, more reliability, and fewer project errors with structured communications.

7. This network of networked teams opens the possibility that working teams can be much smaller and more focused, while not losing the efficiencies of being part of a larger networked team.

8. With networked project teams, a company can maintain centralized control over the critical intellectual capital that resides in project documents—capital that would otherwise remain hidden within the computers of project team members after projects were completed.

SUMMARY

The advent of the cross-functional core team model in the late 1980s and early 1990s revolutionized the performance of project teams and was a primary factor contributing to the significant increase in R&D performance in that generation. The limitation of this model was that it required that core team members be physically colocated, and solely dedicated to one project. Most companies subsequently used email in an effort to overcome the colocation restriction, but this undermined the core team communications model and created unnecessary administrative work for all project team members. By using project network systems instead of email, the networked project team model will revolutionize project team performance.

12 CHAPTER

Enhanced Phase-Review Management

$\mathbf{P}$rior to the Time-to-Market (TTM) Generation of product development management, new product projects seemed to start by some mysterious chain of events, were almost always completed late, and all too frequently failed to realize initial expectations. As one CEO described it at that time, "Our product development pipeline is bloated in the middle like a big bubble where everything sits, and new products don't so much get launched as escape." The introduction of phase-based decision making changed all that.

As good as it was, however, companies eventually discovered some limitations of their phase-review management process. In the new R&D Productivity Generation, these limitations are overcome, increasing the effectiveness of phase-review processes. But before we get into these changes, let's review the principles of phase-review management that we are building upon.

PHASE REVIEWS

Phase-based decision making was one of the most important innovations in the TTM Generation. It enabled companies to invest in new products on a phase-by-phase basis, and to cancel inappropriate or ill-timed projects much earlier. As previously mentioned, some companies saved as much as 10 percent of their R&D budgets through these timely project cancellations alone.

A phase-based decision process requires business approvals of a project at each phase of its development, based on specific criteria for completion of the previous phase and preparedness to enter the next phase. In most companies, the authority for making these "go/no-go" decisions was clarified with the establishment of a new cross-functional executive group, referred to in the PACE® terminology generically as the Product Approval Committee (PAC), but sometimes referred to by other names. A new management process was instituted as well: the phase-review process.

Product Approval Committee (PAC)

Executive responsibility for product development resides with a PAC, but in some cases the responsibility resides with a company's executive committee. In either case, the group has specific authority and responsibility to approve and prioritize new product development investments. This entails initiating new product development projects, canceling and reprioritizing projects, ensuring that products being developed fit the company's strategy, and allocating development resources. The PAC uses a formal phase-review management process to exercise its responsibility and authority.

The PAC meets at the completion of each project phase, reviews the project and plan for the next phase, and then typically convenes in closed session to make a decision. In most companies, the PAC cancels or refocuses problematic projects earlier, thereby reducing wasted development while keeping the worthwhile product development projects better focused on the true needs of the market.

Phase-Review Process

All major product development projects are required to go through a common phase-review process, which generally consists of four to six standard development phases that define the scope and expectations for each phase. The PAC approves all projects phase by phase, so approval only gives the project team the authority to continue through the next phase of development.

Once a project is approved, the PAC delegates authority to the cross-functional project core team to execute the project as proposed. In essence, this creates a highly efficient two-tier management organization for product development: the PAC, which empowers and funds the project; and the core team, which executes it. This practice was originally borrowed from the venture capital model for progressively funding start-up companies as they meet clearly defined milestones.

Stage-gate processes are similar to phase-review processes, but many focus more on stage exit criteria, such as deliverables, and less on the requirements for the next phase. Phase-review processes also use standards to establish the cycle time for each phase, and project teams are empowered to complete the phase within that cycle time. Stage-gate processes generally don't use cycle-time standards to create this sense of urgency.

Not all processes referred to as phase-review or stage-gate are the same. Some vary visibly, while others have more subtle differences that may keep them from succeeding. A successful phase-review process has the following characteristics:

- It provides a clear and consistent process for making new product decisions. At the completion of each phase, the PAC determines if the project should be continued, canceled, or refocused. The PAC also resolves any major issues that arise during development phases.

- Phase reviews establish a higher sense of urgency. Project teams are only empowered to complete the next phase within the defined time, and cannot continue working on it until they think they are ready.

- A phase review is not only a review of the work completed, but also an assessment of the plan for the next phase. Each project team presents its project plan at a high level in a consistent format for the PAC to review. If approval is given, then the team must execute the next project phase according to the approved plan.

- At a phase review, the PAC makes certain that the resources required to complete the next project phase are in fact available, and approves the assignment of those resources.

- Upon approval, the PAC establishes certain tolerances ("contract" items or boundary conditions) for the empowerment of the core team. If the team exceeds these tolerances during the phase, an immediate re-review of the project is expected.

- The phase-review process provides an important link between product strategy and the execution of product development to implement that strategy. In many early cases, the PAC developed product strategy in the phase-review process, but some companies have made the necessary improvements to develop product strategy more formally.

Companies whose phase-review processes do not have these characteristics should correct them as part of any effort to implement this new generation of product development. Otherwise, they won't get all the benefits from either the TTM or the R&D Productivity Generation.

Process Limitations

As beneficial as phase-based decision making is, it still has some limitations:

1. Approval to continue to the next phase requires an assessment of the availability of necessary resources to complete that phase. This assessment was informal at best and unreliable in too many cases, because resource assignment information, and therefore availability, were not tracked. Too often, the resources expected and supposedly approved never materialized, and the project could not be completed as promised.

2. Project approval decisions are made individually when a project reaches the end of a phase. While this is infinitely better than the "no decision" past, the lack of visibility of the impact of the project in question on other projects, including future projects, creates problems. For example, three projects might receive Phase 1 approvals, even though, unbeknown to everyone, there were only resources beyond that point to complete one of them. Or a review committee might approve a project, only to find out a month later that a higher-priority project needed those scarce resources.

3. Without a formal product strategy process, many companies were forced to use their phase-review process to set strategy, instead of managing the execution of a previously developed strategy. While here again this is infinitely preferable to not having any strategy, companies increasingly find the need to define product strategy more deliberately instead of as part of a new product approval.

4. When a project was approved, many PACs applied the best practice of setting boundary conditions or tolerances that defined constraints on the project, such as cycle time, gross margin, project budget, and projected revenue, for the core team. If these boundaries were exceeded, the core team had to come back to the PAC for approval. But there was no system for automatically monitoring these out-of-bounds conditions.

5. Finally, the PAC does not have its own information system or reporting capability, so it relies on project teams to provide the information needed, as well as on individual PAC members to recall the key facts used in previous decisions. This lack of independent information leads to frustration with some PACs, and in some cases to poor decisions.

ENHANCED PHASE-REVIEW MANAGEMENT IMPROVEMENTS

Development chain management systems provide some new capabilities for the PAC and expand the capabilities of the phase-review process. DCM systems make resource assignment information available to project managers and the PAC, so that they have visibility of resource availability in making their decisions. In many companies this visibility will change the phase-review process from individual project decisions to decisions made on groups of projects, in order to make resource allocation a more important consideration in these decisions. When this is combined with improvements in the strategic dimension of product development, some companies may want their PAC to focus more on strategy execution than on strategy setting.

> In many companies, new visibility into resource availability will change the phase-review process—from individual project decisions to decisions made on groups of projects.

DCM systems will also provide the PAC with its own information system, enabling it to better manage its work as an approval body. Much of this information is a by-product of information generated elsewhere in the DCM system, but it can be very helpful to the PAC. Another important improvement for the PAC comes from the capability of the DCM system to automatically monitor changes that affect compliance with specified project tolerances, and to automatically inform the project team and, in some cases, the PAC when a project exceeds its boundary conditions.

I refer to this as *enhanced* phase-review management because it improves the capabilities of the PAC in the next generation. Let's look at each of these improvements more closely.

Phase-Review Decisions with Resource Availability Information

With a DCM system, a company's PAC will have access to accurate information on resource availability prior to making a decision about funding the next phase of a project, and will formally incorporate this information into their decision process. They will expect that project managers have done the work ahead of time to make sure that the necessary resources will be available when needed for the phase, and they will verify this resource availability at the phase review. In some cases, a PAC may require a preliminary assessment of resource availability for subsequent project phases as well. Future resource conflicts may not trigger cancellation of the project, but could indicate to the PAC and the project manager that there is a future issue regarding competition for resources. The PAC will now be able to look at alternatives across projects, instead of simply approving projects one at a time.

CRI, our hypothetical company, recently enhanced its phase-review process to exploit the benefits of its new DCM system, and the use of resource information was a big part of this improvement. Anne Miller put together the project plan and resource requirements for the Phase 1 review of the Fast-Food Robot, seeking approval to continue into Phase 2 of the project. We reviewed this plan in Chapter 6, and you may find it useful at this point to refer to Figure 6–3, which depicts the resource plan for the Fast-Food Robot. Anne was able to get all the resources she needed, except for eight software programmers; Ted Johnson, the manager of the software group, forecasted the need for 20 more people than he had available during the summer, and approximately 10 more than were available for the rest of the year. (Figure 6–4 illustrates this situation.) Ted had only five programmers available in total for July and August. Anne said that she could get along with those five for two months, and then take eight for the remainder of the project, but this would create a shortage of programmers for other projects.

In the past, the PAC would simply have filled Anne's request, only to find after approving another project later on that there were no programming resources left. With a DCM system in place, the PAC asked Anne to offer some alternatives for managing around the programming resource constraint. Anne offered several alternatives:

1.   Assign her team the resources available, including the five programmers available now and then another nine when they become available to make up for the earlier shortage. The project would then be on schedule.

2.   Redo the project plan around a constraint of five software programmers. Anne's analysis showed that this would extend the project by almost a year, cost a lot more, and probably lose the early-to-market advantage.

3.   Assign five software developers now, hire three summer interns to work with them, and hire four software development contractors. This would keep the project on schedule, but it would cost $75,000 to $100,000 more for the contractors over the internal cost.

The PAC approved the third alternative, and Anne commented: "With the way we previously did phase reviews, the PAC probably would have approved my plan without any idea of whether the resources were available. There would then be a problem when the next project came along for approval, and I may have lost some of the resources assigned to me. We now have a workable plan for my project, and most likely for the others still to be approved." On the use of contractors, Anne said, "Interestingly, since we went to a full cost for all our resources, the cost of contractors is not that much higher than internal developers, so my project budget isn't penalized that much."

Combined Phase Reviews

As Commercial Robotics learned from experiences similar to the one just described, it is sometimes necessary to combine phase reviews if different projects are competing for resource requirements. The PAC started to schedule quarterly resource-planning meetings, in which the combined resource needs of all projects were reviewed. These meetings were different from the phase reviews for individual projects, which were still held as usual.

These "resource review meetings," as they came to be known, helped the PAC identify upcoming resource constraints so that they could be included in the committee's decision making. The resource reviews let project managers know when they needed to prepare alternative plans to address resource constraints in upcoming phase reviews, similar to those Anne prepared. In addition, the resource review meetings were sometimes used to make minor adjustments to resource assignments if some conflict had arisen.

Project Tolerances

Like many companies using phase-review management, CRI approved the next phase of the project and empowered the project team with the understanding that it needed a new phase review if specific project tolerances were exceeded. In some companies these tolerances are referred to as boundary conditions or project contract items. While this boundary-setting worked well in theory, it frequently fell short in practice, since there was no formal system for monitoring variations from the tolerances. Project teams sometimes inadvertently exceeded the tolerances for their projects, and this sometimes caused major problems.

On one occasion, a project continued for six months after exceeding its gross margin tolerance. The project was eventually canceled, but in the meantime $1.3 million in development resources were wasted. The project manager claimed he wasn't aware that the gross margin overrun was that significant, but he was fired anyway. CRI expects its project managers to know these kinds of things.

CRI's new DCM system implemented automatic monitoring of major project tolerances. Whenever a tolerance was exceeded, the project manager was immediately notified. CRI considered simultaneously notifying the PAC at the same time, but in the end decided that it was more appropriate to give the project manager a little time to react.

At the Phase 1 wrap-up review for the Fast-Food Robot project, the PAC set three project-tolerance criteria for approval of Phase 2:

- *Phase 2 Plan*
  The project plan for Phase 2 could not extend more than a month beyond the nine months approved. If the plan was extended beyond the end of April, a tolerance deviation would be triggered. Since the DCM system maintained the project plan as part of the enterprise project planning and control system, it constantly checked the scheduled completion date against this tolerance.

- *Project Budget*
  The project budget for Phase 2 was $1,290,000. If this was exceeded by more than 10 percent (or $129,000), then a project deviation would be triggered. Project budgets were integrated within the DCM system, and this tolerance limit was set to trigger whenever the revised budget exceeded the approved limit. Since the project team kept a current project budget at all times, this budget was compared against the original budget, and a deviation would trigger a boundary violation well before the budget was actually spent.

- *Product Gross Margin*
  Product gross margin was the most critical factor in CRI's future financial success. The executives knew that the company wouldn't be very profitable with low gross margins, regardless of sales volumes, so all new product development projects had a gross margin tolerance. For the Fast-Food Robot project, the projected gross margin in Year 3 was used as the trigger point. The robot's third-year gross margin is currently projected to be 61 percent, and the minimum trigger is 60 percent for all new products. Since new product financial planning is now highly integrated at CRI, gross margin is automatically derived from planned selling price, and estimated product cost is determined automatically from product design documents. A design change, for example, could instantaneously trigger a deviation from the gross margin tolerance and initiate corrective actions which, if not successful, would trigger a phase review of the problem.

PAC Information System

Development chain management gives the PAC an information system for managing the phase-review process which itself builds on other information already in the DCM system. This greatly improves the productivity and effectiveness of the PAC by getting its members the information they need when they need it, instead of making them waste time chasing down the information they need. This management information system also ensures that everyone on both the project team and the PAC is aligned and using the same information.

The PAC information system at CRI, shown in Figure 12–1, consists of two categories of information. Illustrated on the right is information exclusively for the PAC. This includes a PAC calendar, with all scheduled phase reviews and dates on which phase-review pre-read documents must be

F I G U R E   12–1

PAC Information System at CRI

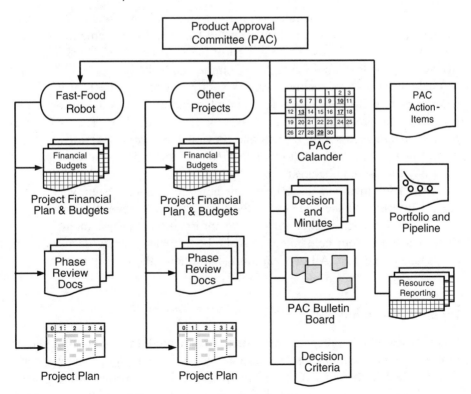

submitted. This calendar is dynamic, meaning that a PAC member simply has to select a meeting date from the calendar, and it automatically links to the documents pertaining to that meeting. If it's an upcoming meeting, then the links are to documents to be reviewed in preparation for the meeting. If it's a previous meeting, then the links are to the minutes for that meeting, as well as to the documents presented at the meeting, a feature which many PAC members find useful.

There is also a PAC bulletin board, where PAC members can post notices and other information, such as articles for other PAC members to read. Live links to other documents can also be posted. In addition, the PAC maintains its own action-item system to keep track of what each member was designated to do following a phase-review meeting. The PAC information system also gives PAC members access to reports from the resource management system.

They have access to a complete set of portfolio and pipeline reports, which they can check on demand. CRI also found it useful to give the PAC access to the standard phase decision criteria and the decisions and minutes from all previous meetings.

The information illustrated on the left-hand side of Figure 12–1 comes from individual projects, and it is accessed by links to the same information used by the project teams. There is a set of information for each active project overseen by the PAC, but the PAC at Commercial Robotics also finds it useful to maintain access to inactive projects as a reference. At CRI, each project team makes available all of its financial planning and budgeting information, its reports and presentations to the PAC, and its project plans. The PAC information system accesses all current and previously approved versions of every project plan, along with each project's financial information, so the PAC can track any trends and changes.

The PAC members at CRI find this new information system very useful. All the information they need is at their fingertips whenever they need it, so they waste far less time chasing down information. Several members also think that the system helps them to make better phase-review decisions than they made in the past. All of them think that it helps them work as a more cohesive team.

Focus of Phase Review

When phase-review processes were first implemented, they exposed a glaring problem at most companies: They didn't have a clear product strategy and therefore lacked the context for making phase-review decisions. Almost inevitably, this discovery led companies to improve their product strategy. In this new generation of product development, some companies carry this improvement even further.

DCM systems will additionally strengthen product strategy and related planning by providing integrated information systems for these processes. Some companies will begin to draw an even sharper distinction between strategy formulation and strategy execution. In some cases, the PAC will become the body responsible for *executing* a new product strategy, while another group will be responsible for *formulating* product strategy. We'll look at integrated product strategy in more detail in Chapter 20.

REQUIREMENTS

Enhanced phase-review management builds on the two previous practices: enterprise project planning and control, and network project teams. The information used by the PAC, as well as the ability to control tolerances, are by-products of the use of the DCM system by all project teams. This means that all project teams need to be actively on the DCM system for the PAC to have visibility.

In addition, an information system needs to be created for the PAC. If all the information is available already, then this is relatively simple. If it isn't, then some additional work needs to be done, which should probably be done anyway.

BENEFITS

The phase-review management introduced in the TTM Generation had some significant benefits, and now the further enhancement of phase-review management enables companies to further reduce wasted development, increase the productivity and effectiveness of their PACs, and improve the execution of product strategy:

1. Automatic monitoring of project tolerances immediately identifies deviations that need attention and raises their importance, avoiding costly delays in refocusing projects.

2. The phase-review information system enables the PAC to be more productive and effective.

3. Resource-based phase-review decisions improve the allocation of scarce resources and avoid nonproductive shifting of resources from project to project.

SUMMARY

Phase-review management was critical to many of the improvements of the Time-to-Market Generation, and continues to be the foundation for the next generation of R&D Productivity. Phase-review management will be enhanced, improving the processes themselves and the work of the PAC. Using its own information systems, the PAC will be able to more accurately assess resource availability in its decisions, and to automatically monitor projects' adherence to approved boundaries.

13
CHAPTER

Integrated Financial Planning and Project Budgeting for New Products

Let's face it: Financial planning and budgeting are generally not exciting to most project managers. These tasks are frequently seen as necessary administrative work, and sometimes mistakenly seen as work that adds little value. Yet financial results are how companies are measured, and, in the end, this will be how the success of a new product is determined. Was the new product worth the investment to develop it?

In the previous generation of product development, financial planning and budgeting were limited to individual spreadsheets prepared by project managers or their financial team members. These plans and budgets were disconnected from underlying operational data and had to be manually updated on a regular basis, or just left to drift away from what was really happening. At the same time, chief financial officers became increasingly unhappy with the state of financial control of R&D in their companies, beginning with the failure to collect accurate financial planning and budgeting information from product development projects.

In the next generation of product development, new management practices significantly improve financial planning and project budgeting for new products. First, let me clarify how I am using these terms. *Financial planning* is the projection of the expected financial results from a new product. *Project budgeting* is the expected investment to develop that product. While they are related, their differences warrant looking at them separately. The horizon for new product financial planning begins after the launch of the product, and it is concerned with measuring the benefit of the product in terms of revenue and profit. The horizon for project budgeting is the development period, and it is concerned with how much is expected to be invested in the product's development.

These new financial management practices are based on the integration of financial information on a new product with related product development information, and with financial information across all projects. New financial planning tools, as part of DCM systems, enable project managers to integrate this information seamlessly, increasing the effectiveness of their financial planning and budgeting. Moreover, these new practices are being introduced at a time when the financial management of product development is becoming more important.

Why is financial planning and budgeting for new products increasingly important? In the mid- and late 1990s, new product opportunities appeared to be numerous. There was little emphasis on how much it cost to develop a product, because the overriding opportunity lay in just getting it to market. There was also little emphasis on the projected financial returns from new products, because the opportunities for innovation were so large and because of the constant threat that competitors would be first to market. As a consequence, many companies paid little attention to the financial planning for new products. But this is no longer the case. It's now increasingly important to ascertain whether the expected financial return on every development investment is sufficient, and whether the calculations of expected returns are reliable. Increasingly, businesses must be able to plan, and to depend on, the financial results of new product development.

> Underestimating the real cost of development is one of the reasons some companies are not getting a good return on their investment in new products.

Understanding the financial return from new products is also increasingly important in prioritizing development opportunities. As R&D investment becomes more limited, companies need to make every development dollar count. At the same time, companies are becoming more aware that the real cost of developing new products is much greater than what they previously assumed, raising the threshold of what is expected from each new investment and increasing the emphasis on financial planning and analysis. Underestimating the real cost of development is one of the reasons why some companies are not getting a good return on their investment in new products.

Chief financial officers and many boards of directors are turning their attention to the financial management of new product development. Many companies invest a lot to develop new products, and their future financial success depends on these investments, so this attention is wholly warranted. Fortunately, this new generation of product development provides CFOs with the new management practices and systems necessary to improve the financial planning, analysis, and control for new products.

Prior to DCM systems, spreadsheets were the primary tools for new product financial planning and project budget management. While it was a start, spreadsheets are very limited. Today, a company wouldn't consider using a spreadsheet to manage inventory, do its accounting, or run its materials planning. Spreadsheets were initially used to perform those very tasks, but enterprise applications eventually took over, raising these business functions to a higher performance level and enabling more advanced management practices. We will see this same transition from spreadsheet to financial system as we go through this chapter.

We'll look at integrated financial planning for a new product, and then at integrated project budgeting. We'll then examine how project budgets are managed as projects progress. Throughout, we'll use the Fast-Food Robot project at CRI to illustrate how these tasks are done differently in an integrated environment. But first, let's review some of the ways in which the integration of financial information will be achieved at the project level.

INTEGRATION OF FINANCIAL INFORMATION

Development chain management systems enable the integration of financial information within a product development project, across all projects, and with standard financial information. This project-related financial information consists of a number of different financial models that all come together as an integrated model, as illustrated in Figure 13–1. With a development chain management system, these models are generally implemented as automated financial planning tools.

The *revenue projection* model defines the expected revenue for the new product over its lifetime or a predetermined period of time. Typically, this projection is made by estimating the number of units that will be sold, along with the expected selling price. In large companies this projection becomes more complex, because there are different types of revenue: for example, services revenue in addition to product revenue, currency-rate-adjusted revenues for different countries, and revenues based on differential pricing. With integrated financial planning, a company can define standard financial assumptions for

F I G U R E 13–1

Overview of Financial Planning Models for a New Product

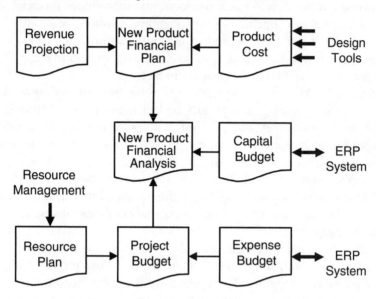

factors such as inflation, time periods, currency exchange rates, and service revenue formulas, so that each team will make consistent revenue projections from a standard set of assumptions.

Early in development, the project team needs to estimate the expected *product cost* of the product being developed, including the materials and labor required to manufacture it. These estimates continually change as the product's design evolves. Early estimates are approximate, but they get more accurate as the product design progresses. In the next generation of product development, the product cost model is integrated with product design tools such as CAD, PDM, CAE, and bill-of-material applications. Product cost can be derived automatically from the product designs as they evolve, and the financial product cost model dynamically updated.

These two models are then brought together with other financial information on the new product to create a *new product financial plan:* in other words, the projected income statement, and in some cases a projected balance sheet, for the new product. This income statement uses the revenue projection and computes the projected cost of goods sold using the product cost model. It then adds projected costs for marketing and sales, administration, customer support, additional development, and so on. In an integrated financial planning process, these cost projections are typically based on standards established by the finance department and applied consistently by all project teams.

The lower portion of the integrated financial model depicted in Figure 13–1 illustrates the budgeting of the costs for developing the new product. The primary costs for most new product development projects are the resource costs required to develop the product, sometimes referred to as the *resource plan*. Integrated financial planning in a development chain management system enables resource costs to be derived automatically from resource management. (Refer to Section Two, especially Chapter 5.) When a project manager changes the resources assigned to the project, the resource costs in the project budget automatically change as a result. In previous generations, these changes were not linked, and the integrity between resources assigned to a project and the costs of those resources was easily broken. As a result, new products frequently cost a lot more than was projected, without any awareness of the impact of the cost overrun on the project's financial return.

In addition to the resource plan, the *expense budget* provides the additional project costs for the project budget. The expense budget includes other development expenses such as travel, purchased research, consulting or contractor costs, services, expendable materials, etc. There are different ways of budgeting these costs, and in the next generation all companies establish a standard approach for each team to follow in managing its expense budget, along with standard financial planning models for implementing the budget. For example, travel expenses could either be budgeted for the project team or allocated from the travel budgets of functional departments. In an integrated financial planning system, functional expense budgets are integrated with project expense budgets.

The *project budget* is the summary of the full cost of developing a new product, combining the resource costs with the other project expenses. With a phase-based decision process, project budgets are approved phase by phase throughout the project. The project budget at the start of every phase becomes the approved version, but project managers may also maintain a current version of the project budget that reflects up-to-date projections at any point in time.

The *capital budget* is used to budget capital expenditures such as tooling, manufacturing equipment, and capitalized software separately from expense items. Capital expenditures are incurred at specific times in the course of the project, but the cost of these items is amortized or depreciated over a period of years. So the timing of capital budget items is different for cash flow and expense.

Finally, these financial models come together in the *new product financial analysis*, which is where the projected profit from the new product is compared to the budgeted cost of developing it. This is where the net profit of the project is computed in the form of return on investment (ROI), net present value (NPV), or other financial measures. In an integrated financial process, this computation is dynamic. Changes in any of the financial models also change the new product financial analysis.

In an integrated financial process, the computation of such measures as ROI and NPV is dynamic.

We've touched on some of the ways that financial information is integrated, so now let's summarize the types of integration. Broadly defined, financial planning and project budgeting for new products are integrated with other aspects of the DCM system in six different ways: standardization achieved through the use of common planning models, project budgets automatically derived from resource assignments, financial data directly based on source data, automatic incorporation of financial information in project documents, project budget updates with actual costs, and consolidation of financial information across all projects.

Common Planning Models

One type of integration is the use of common planning models, such as those we just discussed, in all product development projects. New product financial planning and budgeting are more easily standardized through the use of standard financial planning models for new products. These models are defined by the financial organization within a company, and then easily and consistently applied by every product development team.

In a development chain management system, these financial planning models are sometimes implemented as financial tools that automatically integrate financial information within the broader system. While financial planning models can also be implemented using spreadsheet templates, this will achieve only a portion of the potential benefits. It will not enable automatic integration with other data from resource management or ERP systems; nor will it enable the consolidation of financial data across all projects, since the spreadsheets are usually done and maintained on individual desktop computers. In addition, the deployment and updating of individual spreadsheets to all project teams is difficult, frequently resulting in only partial adoption of standard practices.

Project Budgets Automatically Derived from Resource Assignments

Project budgeting is highly integrated with resource assignments and needs. In a development chain management system, when a developer is assigned to a project, or the project manager identifies a resource need, the cost of that resource is automatically incorporated into the project budget. When the project plan changes, altering the project's resource needs, the project budget automatically changes as well.

The alignment of project budgets with resource assignments and needs is necessary to maintain financial integrity. All too often, a project team puts together a project budget based on estimated resource costs that are not derived from resource needs or assignments, but from separate estimates. When the budget is not derived from assignment information, it is very difficult and time-consuming to manually compute the resource costs for the budget, and, more importantly, it's very time-consuming to manually keep the budget aligned with resource costs throughout the project. As a result, most project teams don't keep them aligned, and when resource needs change (typically upward), the budgets don't reflect the cost changes. For example, $1 million was budgeted for the development of a new product with an expected profit, after paying for the development costs, of $200,000, for a 20 percent ROI. As the project evolved, the team required 15 percent more resources, but the costs were not rebudgeted. If they had been, the financial return calculation would obviously have changed. The new cost would be $1.15 million, and the new profit would be only $50,000, for a return on investment of only 4 percent [$50,000 / $1,150,000]. With an integrated financial planning system, the financial implications of resource assignment changes are directly reflected in the financial analysis for the new product.

Financial Data Directly Based on Source Data

Much of the data used in financial planning for a new product is derived from other sources, and integration directly with the source data enables the financial planning to always be current. We just discussed integration with resource management source data, and the integration with product information is the same for product cost information. Sophisticated design and product data-management tools can automatically compute the cost of modules and components of a product. As the product design advances, there will be a bill of material that can be the basis for setting the product's cost.

Using an advanced development chain management system, a company will link its product design tools directly to its financial models, so that changes to the product design will be automatically transferred to the financial model. Then, when there is a product design change, it is immediately reflected in the financial models. Typically, this link will be created for each product module, since the design application software may be different for each module. The financial model then integrates all of the financial information from the various design systems.

Integrated financial planning for new products also includes the automatic integration of standard financial data, such as assumptions for cost allocations, currency conversion rates, inflation rates, and the like. By integrating this standard financial data into all financial planning models, a company can assure consistency of financial analysis across all projects. Here again, simply publishing these assumptions periodically and expecting all teams to follow them and update their financial plans accordingly is not as effective as automatically integrating the assumptions into the financial models.

Incorporation of Financial Information in Project Documents

Coordinating financial information contained in various documents such as presentations, financial summaries, project plans, and status reports is a constant problem for project teams. At one time or another, everyone has experienced the problems of having financial information that is inconsistent or out of date. In some cases this is simply embarrassing, but in others it may lead to incorrect decisions.

In some development chain management systems, financial information for a project is treated as a data object and might be automatically incorporated into all project documents that use that information. For example, data from a revenue forecast may be incorporated into a new product business plan, an upcoming management presentation, a summary of new product revenue, and the financial analysis for the project. When the revenue forecast is revised, all of the documents using it are automatically revised as well.

Project Budget Updates with Actual Costs

As a project progresses, it incurs actual costs against the budgeted costs, and it's an important function of project management to monitor these actual costs in order to initiate corrective actions when necessary. Since this is very difficult to do when the project budgets are not integrated with the accounting systems (ERP systems), it is frequently not done, or is done too late to be of any use.

Financial planning within a development chain management system can be integrated with an ERP system to collect actual expenses on a regular basis and compare these to the budgeted expenses. As discussed in Chapter 8, actual time can also be collected against assigned time, and this can also be used to automatically track the cost of variances from assigned resources.

Consolidation of Financial Information across Projects

Finally, with an integrated development chain management system, project-level financial information is consolidated across all projects to provide R&D executives and financial managers with the necessary view of product development to make the decisions needed. They are now able, for the first time, to see consolidated revenue forecasts across all projects, as well as the consolidation of capital requirements for new products, and consolidated profit expectations from new products.

This consolidation is illustrated in Figure 13–2. Revenue projections for a new product are automatically combined with and consolidated with revenue projections from other new products to provide a complete picture of projected revenue from all products. Similarly, capital budgets can be consolidated across all projects to provide a unified view of capital requirements, and expense budgets can be consolidated to provide a unified view of expense budgets. Information such as manufacturing volumes, planned direct labor hours, etc., can be consolidated across all projects to provide a complete picture of expected manufacturing requirements from new products.

In general, this consolidation is very difficult when done from individual spreadsheets, because they reside on individual desktop computers. Consolidation requires the periodic collection of all spreadsheets (which can be very time-consuming), adjustment of inconsistencies prior to consolidation, and

F I G U R E   13–2

Financial Information Consolidated Across All Projects

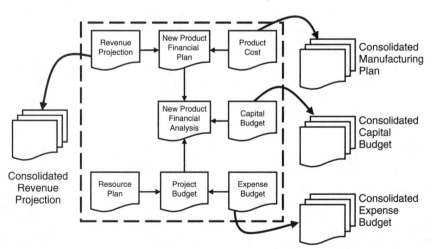

then programming to create the totals across all projects. This is a lot of work that needs to be done each time a consolidation is needed, and given the dynamic nature of product development, the consolidation is usually out of date before it is even completed. With the integrated financial planning capability of a development chain management system, this consolidation is fully automatic and always up to date.

INTEGRATED FINANCIAL PLANNING AT CRI

The best way to illustrate how integrated financial planning actually works is to look at it in action. To do so, let's look at the financial planning for the Fast-Food Robot at CRI, our hypothetical company. Shaun Smith is the company's new CFO, brought in by its board of directors to strengthen financial leadership, and better financial management of R&D is one of his top goals. Shaun summarized his objective in this regard in a presentation to the board of directors shortly after joining the company:

1. Financial planning for new products has been inconsistent and error-prone, and this has caused us to waste precious resources developing products that were not financially justified. To correct this, we will apply standard financial models to all projects.

2. Project budgets have been inaccurate, sometimes based on incorrect data, and not updated as the projects changed, leading to "hidden cost overruns" and a financially out-of-control environment. To correct this, we will implement a standard and integrated project budgeting process.

3. Because of lack of integration, revenue forecasts in the long-range plans have not been based on the approved plans for new products, and functional budgets have not been reconciled with project budgets, leading to severe strategic misalignment. To correct this, all strategic plans and budgets will be based on the consolidation of project-level financial planning and budgeting.

Revenue Projection

Anne Miller, Art Hall, and Richard Salisbury used CRI's new financial planning model to prepare the revenue projection for the Fast-Food Robot, as illustrated in Figure 13–3. Using the new model enabled them to focus all their effort on making the estimates, instead of wasting time on the format and structure for the projections. CRI expected each new product revenue projec-

Fast-Food Robot Revenue Projection at CRI

Fast-Food Robot Revenue Forecast ($000)

	Year 1	Year 2	Year 3	Year 4	Year 5
Americas					
Units	100	500	1,500	4,000	5,000
Cum. Units	100	600	2,100	6,100	11,100
Average Selling Price	$75	$75	$77	$80	$80
Product Revenue	$7,500	$37,500	$115,500	$320,000	$400,000
Installation Revenue	$500	$2,500	$7,500	$20,000	$25,000
Service Revenue		$4,500	$15,750	$45,750	$83,250
Europe					
Units	0	0	300	1,000	1,000
Cum. Units	0	0	300	1,300	2,300
Average Selling Price	$75	$75	$77	$80	$80
Product Revenue	$0	$0	$23,100	$80,000	$80,000
Installation Revenue	$0	$0	$1,500	$5,000	$5,000
Service Revenue		$0	$2,250	$9,750	$17,250
Asia					
Units	0	0	0	500	1,000
Cum. Units	0	0	0	500	1,500
Average Selling Price	$75	$75	$77	$80	$80
Product Revenue	$0	$0	$0	$40,000	$80,000
Installation Revenue	$0	$0	$0	$2,500	$5,000
Service Revenue				$3,750	$11,250
Product Revenue	$7,500	$37,500	$138,600	$440,000	$560,000
Installation Revenue	$500	$2,500	$9,000	$27,500	$35,000
Service Revenue	$0	$4,500	$18,000	$59,250	$111,750
Total Revenue	$8,000	$44,500	$165,600	$526,750	$706,750

tion to use this model, which started with unit sales, identified the expected average selling price (ASP), made sales estimates by geographic region, and projected product and service revenue. The model automatically converted selling prices in foreign currency to U.S. dollars, and did all of the necessary calculations. CRI uses a standard of a five-year projection for all new products, believing that projections beyond five years are not reasonable because of rapid changes in technology.

The Fast-Food Robot Market Study recommended a selling price in the range of $60,000 to $80,000 depending on features, so Anne, Art, and Richard tentatively decided on a $75,000 price for the projection. As the project progressed, they knew that they would fine-tune this price as well as the projected

number of units. The market for the Fast-Food Robot was primarily defined by
the three largest fast-food companies:

1. **McDonald's** is the world's largest fast-food company, with more
   than 30,000 restaurants in 120 countries. Almost 30 percent of its locations are company-owned, with the others run by franchisees. The
   initial pilot locations for the Fast-Food Robot were all expected to
   be company-owned, with the franchisees purchasing the robots after
   they were proven mechanically and financially. There are also
   quick-service kiosk units located in airports and retail areas, but
   these were not expected to be early customers. McDonald's has revenues of approximately $15 billion and more than 400,000 employ-
   ees. The Fast-Food Robots were expected to replace more than
   50,000 of these employees. Each McDonald's unit gets its food and
   packaging from approved suppliers and uses standardized procedures, and it was expected that the company would need to approve
   the models of Fast-Food Robots used in all locations.

2. **Burger King** is the second-largest hamburger chain, with nearly
   11,500 restaurants located in 60 countries. Its revenue is approximately $1.7 billion, and it has more than 30,000 employees. Burger
   King was an early and strong advocate for the Fast-Food Robots and
   was eager to get a jump on McDonald's because it expected that
   customers would initially rush in to see the robots in action.

3. **Wendy's** has about 6,300 fast-food restaurants operating across the
   U.S. and in 21 other countries, and about 80 percent of its locations
   are franchised. The company's revenue is more than $2.5 billion,
   and it has more than 45,000 employees. Wendy's was lagging behind the other two companies in its interest, but did plan on pi-
   lot-testing some robots.

As you can see, the potential market for the Fast-Food Robot was enormous, but the team believed that it needed to base its sales projections on a slow
and deliberate ramp-up. With approximately 50,000 restaurants among the three
chains, and assuming two robots per location on average, the total theoretical
market was 100,000 robots. The team projected that the market would be approximately 25 percent penetrated in five years, and that CRI would have half of
the market, so approximately 12,500 units would be sold in five years. The team
assumed a slow ramp-up in the projections for the first few years.

When they completed this projection, it was automatically linked to
many other systems and documents. It was automatically integrated into the
consolidated revenue forecast for new products and became part of the consol-

idated long-term sales systems. Anne Miller also linked this forecast directly into the team's planning documents, including the Phase 1 project plan and the Phase 1 presentation slides. Anne expected to revise these revenue projections periodically, and now knew that, when she did so, these other systems and documents would be updated automatically, so there would be complete alignment of all the projections.

Anne also thought that the new integrated financial planning model made this task much easier. "Previously, I had to spend a lot of time just formatting the spreadsheet and checking the calculations. I remember one presentation another project manager made where the revenue projections were not added correctly, and instead of a $150 million revenue projection, it was less than $100 million. This was actually pointed out well into his presentation by the CFO, who checked the math with his calculator. The project was canceled and the project manager never recovered. You would be surprised how many simple mistakes like that actually happen, but now, using this new model, I don't need to worry about it."

"I also don't need to do things like tracking down currency-rate conversion projections and updating my estimates. Currency-rate projections are now built into the model. The financial group updates one master currency table, and the projections in all the projects are automatically updated. As a global business, little things like this can be a great help."

Product Cost Estimate

At CRI, the Fast-Food Robot team was the first to fully integrate product-costing with the underlying design tools and systems. As illustrated in Figure 13–4, the team built its product cost model using CRI's standard product cost model, which incorporates the estimated cost of each mechanical and electronic module, as well as the estimated assembly and test costs. The product cost model references the estimated number of units to be manufactured, which differs from the number of units sold by the amount of inventory.

The servo mechanism for the robot was estimated to cost $3,500 to build, based on the mechanical design (Module Design number 30005) for that module. The design was changed recently (a day ago) to incorporate additional body movement requirements, increasing the cost by $400. The cost for this servo mechanism is automatically derived from the design, and as soon as it was changed, the cost estimate in the model was automatically changed. The body design is already developed to the point of a complete bill of material, and the cost is automatically derived from the cost in the bill of material (hence the reference BOM in the source notation).

The Fast-Food Robot head is being designed by an outside company, Walt Disney's Animatronics Division, which was recently created to commer-

F I G U R E   13–4

CRI's Fast-Food Robot Cost Estimate

	Year 1	Year 2	Year 3	Year 4	Year 5	Source of Cost Estimate
Units Manufactured	120	600	2,100	6,300	8,000	Revenue Projection
Mechanical Modules						
Head	$6,500	$6,500	$6,500	$6,500	$6,500	Disney Estimate and Drawings
Servo Mechanism	$3,500	$3,500	$3,500	$3,500	$3,500	Module Design 30005
Body	$2,800	$2,800	$2,800	$2,800	$2,800	Module Design 30026 (BOM)
Arms/Fixtures	$1,500	$1,500	$1,500	$1,500	$1,500	Module Design 30057
Base	$950	$950	$950	$950	$950	Module Design 30034
Other	$500	$500	$500	$500	$500	Miscellaneous
Total	$15,750	$15,750	$15,750	$15,750	$15,750	
Electronic Modules						
Voice Recognition	$1,100	$1,100	$1,100	$1,100	$1,100	Electronic Module #216
Speech	$765	$765	$765	$765	$765	Electronic Module #225
Central Processing	$950	$950	$950	$950	$950	Electronic Module #245
Primary Controls	$850	$850	$850	$850	$850	Electronic Module #301
Interface	$350	$350	$350	$350	$350	Electronic Module #302
Other Electronics	$200	$200	$200	$200	$200	Miscellaneous Estimate
Total	$4,215	$4,215	$4,215	$4,215	$4,215	
Assembly and Test						
Assembly Hours	150	100	75	50	50	Mfg. Planning Worksheet
Cost per Hour	$45	$45	$45	$50	$50	Finance Estimate
Assembly Cost	$6,750	$4,500	$3,375	$2,500	$2,500	
Test Hours	100	75	50	25	25	Mfg. Planning Worksheet
Cost per Hour	$75	$75	$75	$80	$80	Finance Estimate
Test Cost	$7,500	$5,625	$3,750	$2,000	$2,000	
Total	$14,250	$10,125	$7,125	$4,500	$4,500	
Total Product Cost	$34,215	$30,090	$27,090	$24,465	$24,465	

cialize the capabilities of Disney's Imagineers. The Fast-Food Robot team thought that it was critical that the robots look as human as possible, and turned to the company they thought was the best at achieving this effect. While an expensive part of the development, it looked as though this partnership would give CRI a very big advantage over its competitors, who were also trying to develop fast-food robots. The cost estimate in the product cost model automatically comes from the external link to the Disney design group doing the codevelopment. The group is continually changing the design, and the cost estimates show up in real time in the cost model. For example, a few hours ago the estimate changed from $6,575 to $6,500.

The product cost for the Fast-Food Robot is currently estimated at $34,215 in the first year, then declining in subsequent years. This version of the cost changes continually, enabling the team to know exactly what the cost is at all times. CRI also "freezes" the product cost at the end of every phase for the phase approval, and then monitors changes from this baseline. If the "real time" cost estimate exceeds project tolerances, the DCM system automatically triggers a special phase review. Additionally, Anne's team tracks the trends for the product cost using a monthly chart produced for this purpose.

New Product Financial Plan

The new product financial plan is the summary of the financial expectations for the new product. The Fast-Food Robot team prepared the projected income statement illustrated in Figure 13–5.

F I G U R E   13–5

CRI's Fast-Food Robot New Product Financial Plan

Fast-Food Robot Financial Plan ($000)

	Year 1	Year 2	Year 3	Year 4	Year 5
Units Sold	100	500	1,800	5,500	7,000
Cum. Units	100	600	2400	7900	14900
Revenue					
Product Revenue	$7,500	$37,500	$138,600	$440,000	$560,000
Installation Revenue	$500	$2,500	$9,000	$27,500	$35,000
Service Revenue	$0	$4,500	$18,000	$59,250	$111,750
Total Revenue	$8,000	$44,500	$165,600	$526,750	$706,750
Cost of Sales					
Product Cost	$3,422	$15,045	$48,762	$134,558	$171,255
Installation Cost	$300	$1,500	$5,400	$16,500	$21,000
Service Cost	$0	$2,700	$10,800	$35,550	$67,050
Depreciation					
Total Cost of Sales	$3,722	$19,245	$64,962	$186,608	$259,305
Gross Margin	$4,279	$25,255	$100,638	$340,143	$447,445
(GM %)	53%	57%	61%	65%	63%
SG&A					
Selling and Marketing (40%)	$3,200	$17,800	$66,240	$210,700	$282,700
Administration (10%)	$800	$4,450	$16,560	$52,675	$70,675
Continuing R&D	$1,000	$1,000	$3,000	$3,500	$4,000
Total SG&A	$5,000	$23,250	$85,800	$266,875	$357,375
Net Income	-$722	$2,005	$14,838	$73,268	$90,070
(NI %)	-9%	5%	9%	14%	13%

The team used the standard model maintained by CRI's finance group, and most of the information for the new product financial plan came from data included in other financial models, which are integrated here. Revenue was automatically included from the revenue projection, and the product's cost of sales was computed based on the product cost multiplied by the projected unit sales. Installation cost and service cost were automatically computed based on percentages maintained by Finance and applied consistently to all projects, so the team didn't need to compute them. CRI's practice is that a team could propose an exception to these percentages, but it would then have to submit the proposed exception for phase-review approval. These percentages are updated periodically to reflect CRI's current costs, and when they are updated all financial plans are immediately changed to reflect the new percentages.

Similarly, selling and marketing expense and administration expense were automatically computed at 40 percent and 10 percent of revenue, respectively, based on standards from Finance. These are also changed from time to time, and all new product financial plans are immediately updated.

All new product financial plans at CRI are expected to include estimates for continuing R&D. Since the Fast-Food Robot is the first release of a new type of product, the team expected that a continuing investment would be needed to add features, improve usability, and correct problems, so they budgeted for some continuing R&D.

The plan shows that the new product would be profitable in Year 2, and profit would increase to $90 million by Year 5.

INTEGRATED PROJECT BUDGETING AT CRI

As defined in Figure 13–1, a project budget consists of several financial models: the resource cost model, the project expense budget, and the project budget. These all come together at CRI, along with the financial information for the product plan, as an integrated financial analysis.

Development Resource Cost

Resource costs are usually the most significant costs in a project budget. In a fully integrated development chain, the costs associated with project resource assignments and needs are automatically computed from these assignments. When a project manager or resource group manager assigns someone to a project, or when project needs are determined by the project manager, the project budget at CRI is automatically updated with these costs. When resource assignments or needs change, the cost impact of these changes is immediately reflected in the project budget, so that the information in the resource management system and project budgets at CRI is always aligned.

Development resource costs are generally budgeted based on the extent of the assignment and the cost of the resource. Resource costs were discussed in Chapter 5, and it might be useful to go back and review that discussion. One of the benefits of integration is that resource costs are established as a standard that is used consistently across all projects. The resource costs for the Fast-Food Robot project budget at CRI were computed directly from the project's assignments and needs, which we previously discussed.

> Resource costs are usually the most significant costs in a project budget. In a fully integrated development chain, the costs associated with project resource assignments and needs are automatically computed from these assignments.

Expense Budget

In addition to resource costs, a project also needs to establish a budget for expenses. These include routine expenses, such as travel and market research, as well as significant expenses, such as contractors, consulting, and materials for prototypes.

Sometimes, expenses are already contained in functional budgets and need to be "allocated" to project budgets. An example of this is travel budgeted by individual departments. When a project team budgets travel, it needs to budget it by expense and department, such as marketing travel or engineering travel. This requires additional detail for project budgeting, and some companies have begun to budget these expenses by project instead of department. This was what CRI did.

Project Budget

The project budget consolidates resource costs and project expenses, as is illustrated in Figure 13–6 for the Fast-Food Robot project. At CRI, all projects are budgeted by phase, since the project budgets are approved by phase and actual costs are tracked by phase. Within a phase, the budgets are also broken down by month so that budgets can be consolidated by month for financial planning and actual costs can be compared monthly. But the phase budget is illustrated here, since it is the most significant budget view at CRI.

The budgeted resource costs for this project are derived automatically from the project's resource assignments and needs. The cost of software devel-

F I G U R E 13-6

CRI's Fast-Food Robot New Project Budget

	0 Actual	1 Budget	2 Budget	3 Budget	4 Budget	Total
Project Resource Budget						
Project Management	$56.0	$90.0	$135.0	$45.0	$30.0	$356.0
Marketing	$75.0	$67.5	$67.5	$22.5	$75.0	$307.5
Software	$45.0	$135.0	$1,350.0	$360.0	$30.0	$1,920.0
Finance	$35.0	$90.0	$48.8	$26.3	$7.5	$207.5
Electrical Engineering	$25.5	$67.5	$675.0	$165.0	$30.0	$963.0
Mechanical Engineering	$54.3	$127.5	$502.5	$90.0	$30.0	$804.3
Quality	$18.0	$56.3	$86.3	$135.0	$15.0	$310.5
Manufacturing	$12.0		$142.5	$45.0	$45.0	$244.5
Test	$10.0		$112.5	$202.5	$22.5	$347.5
Other	$55.0	$50.0	$150.0	$150.0	$100.0	$505.0
Total	$385.8	$683.8	$3,270.0	$1,241.3	$385.0	$5,965.8
Expense Budget						
Travel	$35.2	$75.0	$110.0	$75.0	$20.0	$315.2
Market Research	$125.7	$350.0	$0.0	$0.0	$0.0	$475.7
Software Expense	$7.6	$10.0	$250.0	$50.0	$0.0	$317.6
Contract Design	$0.0	$0.0	$1,250.0	$250.0	$0.0	$1,500.0
Materials Expense	$0.0	$0.0	$350.0	$75.0	$0.0	$425.0
Consulting	$125.5	$100.0	$50.0	$25.0	$50.0	$350.5
Other	$12.5	$75.0	$100.0	$75.0	$50.0	$312.5
Total	$306.5	$610.0	$2,110.0	$550.0	$120.0	$3,696.5
Allocated Development Costs						
Voice-Processing Technology					$2,500.0	$2,500.0
Mechanical Systems					$1,200.0	$1,200.0
Other					$500.0	$500.0
Total	$0.0	$0.0	$0.0	$0.0	$4,200.0	$4,200.0
Project Total	$692.3	$1,293.8	$5,380.0	$1,791.3	$4,705.0	$13,862.3

opment is expected to be a third of the resource cost of the project, and most of the resource costs are expected to be incurred in Phase 2. (Both expectations are typical for a project of this type.) The Fast-Food Robot project is currently at the end of Phase 0, so those costs are actual, and the remaining phases are budgeted.

The expense budget at CRI can be broken down in more detail, but it is shown here with summary categories. The total budget for this project is expected to be almost $14 million, including allocated development costs. Within the expense budget is $1.5 million budgeted for contract design. Most of this, $1.3 million, relates to the contract costs for Walt Disney's Animatronics Division to design the head and upper body of the robot. Most of these costs will be incurred in Phase 2.

When expenses are budgeted at CRI, the budget line item identifies the general ledger account number, which enables integration with CRI's accounting system. Actual expenses are collected by account number and the project code for the Fast-Food Robot so that they can be compared to the project budget. CRI recently enabled this by integrating its ERP system with its DCM system.

The Fast-Food Robot is part of a broader program, the Retail Robot Program, and utilizes some critical technology developed for that program. CRI expects that all projects using common technology will share the cost of developing that technology. These shared costs are budgeted as Allocated Development Costs. The primary cost is for the voice-processing technology to recognize and interpret a wide variety of human voice characteristics and speech patterns, which is one of CRI's advanced technologies and competitive advantages.

Capital Budget

Capital expenses are sometimes budgeted separately, because the cash flow and expense for capital items have different timing. For example, tooling required to produce a new product needs to be created and paid for during product development, but the cost of that tooling is amortized over the period that the tool is used to produce the product.

CRI puts the same item in two different financial models: the capital budget contains the amounts necessary for the capital spending in the first place, but the amortized cost is included in manufacturing overhead used in the product cost model.

Project Financial Analysis

The project financial analysis combines the financial planning for a new product and the project budget to develop the product in order to determine the financial viability of the opportunity. There are several measures for this determination, such as return on investment, net present value, and internal rate of return, as well as time to profitability or break-even point. Generally, these are all correlated, but some companies may have a preference for one over another. Some companies have a very specific hurdle rate, which may vary by the type or risk of project, while others may only use these financial measures as a guide.

The Fast-Food Robot project team applied the standard financial model required by the CFO. Even though this was the first time many of the team members had used this new model, Art Hall, the financial manager on the project, was experienced with it since he helped to create it. To the surprise of the team, they didn't need to do any work at all, since the data were automatically derived from other financial models for this project. The financial analysis is illustrated in Figure 13–7.

The model starts with the planned income or loss for each year from the Fast-Food Robot financial plan. The budget for each project category is automatically derived from the overall project budget, with the phase budgets allocated proportionally by year since the financial analysis is done by year. The model totaled these budgets to determine the net income or expense by year, showing a cumulative income from the project of $170 million before taxes. The financial analysis model also automatically computes the cash flow for the

F I G U R E   13–7

Financial Analysis for the Fast-Food Robot Project at CRI

Fast-Food Robot Project Financial Analysis

	Year 2 Phase 0	Year 1 Phase 1	Phase 2	Year 0 Phase 2	Phase 3	Phase 4	Year 1	Year 2	Year 3	Year 4	Year 5	Rem.
Planned Income (Loss)							($722)	$2,005	$14,838	$73,268	$90,070	
Project Budget												
Resource Cost	$386	$684	$2,191	$1,079	$1,241	$385						
Expenses	$307	$610	$1,414	$696	$550	$120						
Allocated	$0	$0	$0	$0	$0	$4,200						
Total	$692	$1,294	$3,605	$1,775	$1,791	$4,705						
			$4,898			$8,272						
Net Income (Exp.)	($692)		($3,605)			($4,705)	($722)	$2,005	$14,838	$73,268	$90,070	
Cum. Income (Exp.)	($692)		($4,297)			($9,002)	($9,723)	($7,718)	$7,120	$80,387	$170,457	
Cash Flow												
Capital Budget				$5,000								
Working Capital Inc.							$1,333	$6,083	$21,517	$66,275	$51,517	($51,517)
Net Cash Flow	($692)		($3,605)			($9,705)	($2,055)	($4,078)	($6,679)	$6,993	$38,553	$51,517
Cum. Cash Flow						($14,002)	($16,057)	($20,135)	($26,814)	($19,821)	$18,732	$70,249
Net Present Value	$24,816											
Simple ROI	897%											
Internal Rate of Ret.	31%											

project. Here, the amount for capital spending, $5 million, was added to Phase 4, along with the incremental working capital required for accounts receivable and inventory. The CRI financial analysis model requires all new products to be successful in five years, so it assumes that all remaining working capital, $51 million in this case, is returned as cash flow after five years.

CRI uses three financial measures to evaluate the projected financial success of a new product opportunity. Net present value computes the discounted net cash flow using a discount rate, and in CRI's case, it's 10 percent per year, thereby valuing near-term cash flow greater than longer-term cash flow. Discounted at 10 percent, the Fast-Food Robot project would be expected to result in a net present value of almost $25 million. A simple ROI, which takes the planned income adjusted for income taxes and divides it by the budgeted project cost without consideration of cash flow or the time value of money, results in an 897 percent ROI. The internal rate of return, which computes the return on investment based on cash flow and the time value of money, is 31 percent. CRI uses the internal rate of return as its primary metric, and expects all major projects such as the Fast-Food Robot to have at least a 25 percent internal rate of return (IRR).

Art Hall summarized the benefits of this practice to CRI and the importance of how all of this fits together. "Previously, every team used its own approach to financial projections and budgeting, and because of all the inconsistency we regularly approved some projects that were not justified financially. And I suspect we may also have canceled some other projects for financial reasons that weren't very sound. Now we have a consistent approach to the financial planning, budgeting, and financial analysis of all projects. These financial models are all integrated, so a change in, let's say, the product cost automatically changes the financial projection and shows the impact on the financial analysis in real time. We no longer go for weeks not knowing the impact of a change; we can react immediately. We also automatically integrate this financial information into what we call our project tolerances, which are the conditions for approval by our Product Approval Committee. If the IRR or the projected income or the gross margin changes by a predefined percentage, the team is automatically notified by our development chain management system, and a phase review meeting is initiated."

Art also described how he manages the financial information in project documents. "We just started to embed this financial information in all relevant project documents, such as presentations and business plans. When a change is made to the financial model, the documents using the model are automatically updated as well. This avoids many of the problems we had in the past and ensures that everyone is using the same financial information."

FINANCIAL MANAGEMENT OF NEW PRODUCT PROJECT

Once the financial plans for a new product and the project budget to develop it are approved, then the project manager needs to manage progress to the budget and track revisions to the financial plan to ensure that the project is under sufficient financial control.

With an integrated development chain system, the necessary financial management reporting is done automatically for the project manager. Approaches to this automatic reporting vary across companies, based on the preferences of project managers and the degree of responsibility the CFO takes for standardizing this reporting across all projects. At Commercial Robotics, the new CFO took his responsibility for this very seriously, requiring all project teams to manage project budgets and track changes to the financial plans for the products according to specific reports. He also reviewed many of these budgets himself, and used them as an opportunity to mentor project managers to be better financial managers. "Most of our project managers don't have a financial grounding. That's okay, since we hired them for other reasons, but we expect all of them to properly handle the financial management of new products. We don't expect them to each develop their own financial management process; we've done that for them. It's a standard process, and we train them all in how to interpret and react to the financial information. I also pick one or two projects a month to review, and sit down with the project managers to mentor them. At first, they expected this to be a confrontation, but now I think they see it as a chance to learn."

At CRI, there are two approaches to the financial management for new product projects. The first is managing actual costs against the project budget. The second is managing revisions to the financial plan. The company is also experimenting with using earned value analysis (EVA) on several projects.

Actual-Cost Management

CRI manages actual costs against budget by phase. Other companies may find it better to manage their actual costs on a month-by-month basis, but since CRI approves project budgets on a phase-by-phase basis, actual costs against budget are best managed by phase. An example is illustrated in Figure 13–8.

The actual costs for project resource budgets are automatically updated from the resource management system based on time actually charged to the project. Before CRI implemented actual time tracking, "actual" resource data were based on resource assignments. This approach was sufficiently accurate. Actual expenses are automatically updated from the company's ERP system,

Actual-to-Budget Comparison for the Fast-Food Robot Project at CRI

Fast-Food Robot Phase 1 Actual-to-Budget Comparison

55% Complete	Phase 1 Budget	Actual thru March	Remaining Budget	% of Budget
Project Resource Budget				
Project Management	$90.0	$44.6	$45.4	50%
Marketing	$67.5	$44.3	$23.2	66%
Software	$135.0	$88.7	$46.3	66%
Finance	$90.0	$42.8	$47.2	48%
Electrical Engineering	$67.5	$33.1	$34.4	49%
Mechanical Engineering	$127.5	$95.3	$32.2	75%
Quality	$56.3	$10.2	$46.1	18%
Other	$50.0	$10.0	$40.0	20%
Total	$683.8	$369.0	$314.8	54%
Expense Budget				
Travel	$75.0	$53.1	$21.9	71%
Market Research	$350.0	$205.5	$144.5	59%
Software Expense	$10.0	$3.5	$6.5	35%
Contract Design	$0.0	$0.0	$0.0	
Materials Expense	$0.0	$0.0	$0.0	
Consulting	$100.0	$43.5	$56.5	44%
Other	$75.0	$22.0	$53.0	29%
Total	$610.0	$327.6	$282.4	54%
Project Total	$1,293.8	$696.6	$597.2	54%

which maintains actual expenses in the general ledger and accounts payable systems. When these expenses are entered, they're charged to a project code, the Fast-Food Robot in this case, as well as to the account number.

CRI's actual-to-budget comparison tracks the percent of the budget used so far in each phase of a project. In March, the Fast-Food Robot project was halfway through the elapsed time of Phase 1, and the project team estimated that it was approximately 55 percent complete. Project costs are not always incurred proportionately to the elapsed time or the percentage completion, so comparing the percentage of budget to the percentage complete is only a rough indicator.

CRI requires that the project manager and financial manager review the actual-to-budget comparison every month and summarize any key items in a memo to the CFO, with copies to the Product Approval Committee. The actual-to-budget comparison, along with a memo, are attached to the Fast Food Robot's project management step in the DCM system, and all previous versions are available to everyone with authorized access. Here is the memo for the month of March:

> To: Shaun Smith, CFO
> From: Anne Miller, Art Hall
> Subject: March Actual to Budget for Fast-Food Robot
> We continue to believe that we are closely tracking the project budget overall, even though there are some variances in particular categories:
>
> 1. Most resource costs are tracking according to the assigned time and progress completed (55 percent). Our only concern is Mechanical Engineering, which is taking more time than expected due to some difficult design issues. We are in the process of rebudgeting this in the current budget.
> 2. Travel costs are high because we accelerated some of the customer visits in order to manage our time more efficiently, but we still expect this to come in at budget.
> 3. We have purchased most of the market research we need and expect to have a favorable variance of at least $100K for the phase. We were able to negotiate better discounts.
> 4. The percentage of the overall budget incurred is close to the progress we estimate that we've made so far. When we update the current budget we should still be close to the approved budget.

CRI maintains several different versions of the project budget. "Phase Approved Budgets" are established at each phase approval. These are the official record of what was approved, and each project team tracks the changes in its approved budgets as they progress though all the phases of development.

In addition, each project team maintains a current project budget that contains the current view of the expected costs for both the current phase and the entire project. It is expected that each project team will update this view of anticipated costs at least monthly. When the current project budget exceeds the approved budget by a predetermined percentage or amount, then an exception is automatically triggered by the DCM system and a new phase review is required.

Earned-Value Analysis

CRI is also experimenting with earned value analysis for each of its projects. Earned value was automatically computed for each project, since the percent-

F I G U R E   13–9

Earned Value Analysis for the Fast-Food Robot Project at CRI

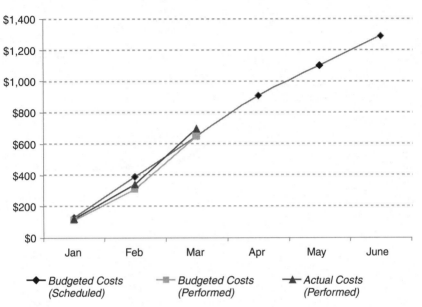

age of completion was managed in the DCM system and the actual costs, as described above, were also computed.

Figure 13–9 illustrates the earned value analysis for the Fast-Food Robot project. As you can see, the budgeted costs scheduled for the six months of Phase 1 are cumulative. The budgeted cost of the actual work performed is based on the budgeted costs for the percentage completion of each step in the phase. It started out a little behind, but has now gone a little ahead. The actual cost of the work performed tracked the delay in progress, but has not caught up.

Revision Control

The other aspect of financial management is tracking revisions to financial estimates. At CRI, key financial items are automatically tracked by the financial management application within its DCM system. As Figure 13–10 illustrates, the original estimates at the beginning of Phase 0 were higher, then came down as the team began working on Phase 0. By the Phase 0 review, the projections were more realistic, and since then the revised estimates have not trended much. There has been some adjustment in pricing and volume esti-

F I G U R E 13-10

Key-Item Trend Report for the Fast-Food Robot Project at CRI

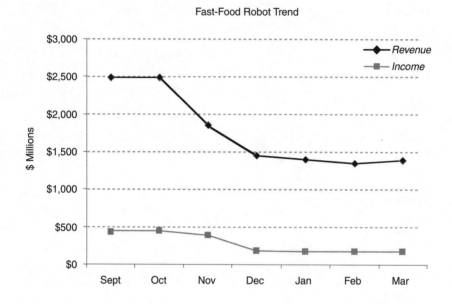

Fast-Food Robot Trend

mates, but the combination of all these changes has had little impact on the overall total estimates.

Shaun Smith talked about how important this trend tracking is at CRI. "Seeing how the estimates of key financial items are tracking helps the project manager see how the expectations are changing. Previously, we only updated these estimates at phase reviews and there were too many surprises—most of them bad. I don't like surprises, especially bad ones."

REQUIREMENTS

These financial management practices are based on an integrated financial system that is part of a broader DCM system. As stated previously, it is very difficult, if not impossible, to do this analysis with spreadsheets.

In addition, standard planning models need to be developed. While this is a significant effort, it generally is recovered in cost savings when these models are used only a few times. In developing these standard models, some additional financial policies may be required, but these are usually necessary for good financial management anyway.

BENEFITS

There are a broad range of benefits from integrated financial planning and project budgeting. Some of these have to do with efficiencies gained by automatically integrating all of the financial information for a project, while others have to do with the efficiencies from standard financial models. But the biggest benefits come from improved financial planning and management. The automatic integration of financial information is so important because most project managers just don't have the time to do the necessary administrative work to collect all of the information manually, so they don't do it effectively.

The following are some of the major benefits that a company can expect from implementing an integrated financial planning and project budgeting process:

> The automatic integration of financial information is so important because most project managers just don't have the time to do the necessary administrative work to collect all of the information manually.

1. Standard planning models, as well as the automatic integration of standard financial planning data, enable consistent financial planning across all projects.

2. Automated financial planning enables project managers to focus on the critical assumptions instead of on the administrative work of putting the numbers together, thereby increasing the reliability of the financial plans.

3. Automatic integration of all financial data ensures that everyone is working from the same information.

4. Automatic updating of actual costs in comparison to budget gives the project manager the information necessary for financially managing his project.

5. The entire project team gets visibility of all financial planning, budget, and variances, which enables them all to understand what is happening.

SUMMARY

We started this chapter by highlighting the increasing importance of financial planning and budgeting for new products, in spite of the reluctance of many project managers to spend the necessary time to do this adequately. We also discussed increasing leadership by CFOs in new product development. Financial planning and project budgeting for new products constitutes an integrated set of models within a development-chain management system, and the integration comes in several ways, including the ability to consolidate financial information across all projects. We discussed how integrated financial planning and budgeting is an automatic result of other operational data, and this makes it both more accurate and easier for project teams. Finally, we illustrated the workings of integrated financial planning and budgeting with a comprehensive example of integrated financial planning and project budgeting for the Fast-Food Robot project at CRI.

14
CHAPTER

Distributed Program Management

Until now, the management of interrelated projects has been cumbersome, and the benefits of coordinating related projects have been elusive. The traditional practice of managing each development project as an independent, isolated entity ignored what were often very important interdependencies among projects.

With the advent of three new capabilities of the next generation of product development that we discussed in previous chapters, it is now possible to seamlessly manage a set of related projects as an integrated whole. With enterprise project planning and control becoming a management process instead of a tool, the planning and control of multiple projects can now be coordinated easily. The discussion of networked project teams introduced the concept of a network of networked teams as a preview of how the work of multiple projects can be coordinated. The third capability is integrated financial planning and project budgeting, enabling the consolidation of financial information across related projects.

At the risk of some confusion, I'm going to use the term "program management" to broadly describe the management of a set of related projects. The term "program" has been used in different ways by different people, and in some cases the terms "project" and "program" are used interchangeably. Nevertheless, there is sufficient use of the term "program" in the context I want to use. *A Guide to the Project Management Body of Knowledge,* for example, defines a program as "a group of related projects managed in a coordinated way."[1] So a program is a set of related projects, and the term *distributed program management* refers to the capability to distribute the management of related projects while simultaneously managing them as an integrated program.

There are various types of related projects in product development. For example, the second version of a product is sometimes developed in parallel with the first version, so that the two can be released in rapid sequence. In this case, some project work obviously needs to be coordinated across the two versions.

A new product could also require the completion of other development projects of a nonproduct nature. Developing a new manufacturing process and creating the channel for distributing the new product are two examples. While these could all be managed together as a single project, there are advantages to managing them as separate but related projects. The three projects will have different time schedules and approval points, will involve very different kinds of work, and will call for project managers with different skills.

Sometimes new technology is developed in parallel with a new product applying that technology. Since technology development is very different from product development, there are advantages to managing the two separately, particularly if the technology is to be used in multiple projects. The same may be true for developing subsystems that are used across multiple new products. Yet at the same time, all of these related projects need to be closely coordinated.

Some companies develop multiple products in a product line that are based on a common platform, and in these cases there are advantages to coordinating the work on each product. The projects are related to the same product-platform elements, and their financial forecasts are usually interdependent. Both the platform and the products in a product line can be managed as a program.

Finally, complex projects are sometimes better managed when broken up into smaller projects, each with its own smaller team and more specific focus. Yet these smaller projects need to be highly integrated throughout, so there is a trade-off between the benefits of increased focus and the overhead and inefficiency of coordinating multiple projects for the same product. With the distributed program management capabilities we will be discussing, the overhead and inefficiency of coordinating multiple projects is greatly reduced, and this balance now shifts in favor of smaller, related projects.

We will look at distributed program management as having three dimensions. The first is coordination of planning and control of multiple related projects within a program. The second is using a network of networked teams to coordinate the related work of multiple project teams within a program, and the third is the financial integration of all projects in a program. But before we examine these dimensions, let's take some time to understand the unique characteristics and program management needs of different types of related projects.

PRODUCT DEVELOPMENT PROGRAMS

As we just discussed, there are different types of product development programs, and each of these has different characteristics and somewhat different objectives for the inter-relationships of the projects in the program. We will use Figure 14–1, an example from the Retail Robot Program at CRI (our hypothetical company), to illustrate each of these program types. At the top of the figure are three primary product development projects: the Fast-Food Robot I, the Fast-Food Robot II, and the Retail Robot. Two module projects (Robot Head; and Arms and Fixtures) within the Fast-Food Robot project are also shown. There are three platform technology development projects: voice

F I G U R E **14–1**

CRI's Retail Robot Program

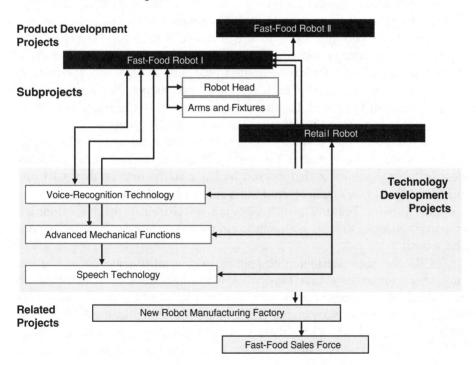

recognition technology, advanced mechanical functions technology, and speech technology. Finally, there are two related projects with a broader scope than the Fast-Food Robot: the new robot-manufacturing facility and the new fast-food sales force. The relationships among each of these projects in the Retail Robot Program are a little different.

Coordinating Multiple Versions of a Product

In some cases, a company will work on multiple versions of the same product in parallel, usually by beginning work on the next version while the previous version is being completed. This tactic allows the company to bring a product to market quickly, in order to establish a "beachhead," and then to quickly follow up a version with enhanced features and capabilities, thereby securing and expanding the beachhead. Such parallel development is typical in software companies. One version of a product may be in the testing stage, while the next version may be in programming. There might even be a third version in the design stage. By following this development strategy, a company can continuously bring new versions of a product to market and effectively apply its development resources to what it does best.

Program management for multiple versions of the same product requires coordinating the product scope across all versions. Product management generally guides the coordination of what is included in each version as part of an overall product strategy and is subject to approval by the company's Product Approval Committee. Under distributed program management for multiple product versions, the scope of each version of the product is tightly shared and coordinated, but there may be little need to coordinate the interaction of the different development teams or synchronize the project plans.

At CRI, the Fast-Food Robot II project is scheduled to start after the design for the Fast-Food Robot I project is finalized, and its scope is to add new features to the robot that weren't included in the first version. CRI uses distributed program management in this case to coordinate the specifications for each version. Features of the robot are prioritized for each version, and important features that are not included in the first release will be included in the second.

CRI also uses distributed program management to consolidate financial projections for each version. Otherwise, it would be double-counting the expected revenue or burdening the first version with all of the development cost. We had a glimpse of program budgeting in the financial plan for the Fast-Food Robot in Figure 13–5, where there was a budget item for continuing R&D. This item is the funding estimate for the Fast-Food Robot II project.

Managing a Complex Project as Subprojects

There are advantages to breaking down a complex project into subprojects, and distributed program management now provides the systems and processes for simultaneously managing these subprojects as individual projects and as a common program. With subprojects, a smaller project team can efficiently focus on developing a module of a product or some other portion of the product that is reasonably distinct. Yet, at the same time, project planning and control needs to be tightly synchronized across all subprojects, and with the primary project. And the subproject teams all need to use the same information, or they risk not being able to bring the product modules together in the end.

Distributed program management now provides the systems and processes for managing subprojects as individual projects and as a common program.

Two subprojects are illustrated in our Commercial Robotics example. The robot head is a subproject being done externally by Disney, and the arms and fixtures for the robot are being managed internally as a subproject. These two subprojects require extensive coordination of project plans and development information.

When used to manage the subprojects of a complex product, distributed program management must tightly synchronize project plans; typically a step in the overall project represents a subproject. The entire program team—subproject teams and the overall project team—needs to use the same information and be able to self-organize around common work whenever necessary. Financially, product plans and budgets are usually unified for a program with subprojects. This is the most complex form of distributed program management, and we'll look at it in more detail in the three dimensions of distributed program management.

Integrating Common Technology Development Projects with Multiple Product Development Projects

Frequently, a company will develop a new technology to be used in multiple products. Traditionally, the new technology may be included in the development project for the initial product, using it in order to get the new product developed quickly, but there are some problems in doing it this way. Technology development is very different from product development; it is more uncertain, follows a different process, and calls for a broader and more farsighted perspective. The lack of this perspective can be costly. Developing a technology with a single product in mind, for example, typically results in the need to redo the technology before it can be used in the next product.

In our CRI example, three platform technology projects are illustrated, along with their relationships to the Fast-Food Robot and Retail Robot projects. These technology projects are managed separately from the product development projects. They have their own project managers, who are skilled at managing research projects. The platform technology projects were started well ahead of the product development projects, since CRI knew that the technology development would take much longer and would be applied to multiple projects. In fact, CRI launched these technology development efforts before it had fully committed to the Fast-Food Robot. Technology development projects also follow a different type of development process, with different phases, approval points, and work flows.

Distributed program management for shared technology development focuses on synchronizing the availability of the technology with the needs of the product team. This entails integrating the technology and product development schedules through enterprise project planning and control, although the two are generally integrated toward the end of the technology development project, which is different from the previous case. Distributed program management also calls for synchronizing the specifications, performance characteristics, and costs of the technology and the product, which requires that the product and technology teams share the same technical information.

Coordinating Related Product and Process Development

Distributed program management can extend beyond complex projects and their underlying technology. When looking comprehensively at the work required to launch a new product, other changes may be necessary, such as changes to the manufacturing process for a new product, the customer support process, the sales and distribution process, and so on. In some cases these changes are simple enough to be managed as steps or even tasks in the primary product development project. In other cases, however, there are benefits to managing these "supporting changes" as separate but related projects with their own teams, project managers, and timing.

In the Commercial Robotics example, a new robot-manufacturing facility is being built, and the Fast-Food Robot is the first product to be manufactured in it. Since this new factory will be used for other new robots as well, and since the factory is such a big project in its own right, CRI is managing its construction as a distinct project, but is using distributed program management to synchronize the two projects. Similarly, CRI has initiated a related project to create a new sales force to sell Fast-Food Robots and related products.

The relationships within distributed program management between the primary product development project and related projects vary, based on the nature of the projects. The manufacturing-plant building project, for example, needs to be tightly synchronized with the robot's development so that the plant will be ready in time to build the robot and will be able to meet the projected production needs. Yet at the same time, CRI doesn't want to build the factory too early, only to have it sit idle.

DISTRIBUTED PROGRAM PLANNING AND CONTROL

Enterprise project planning and control (both the system and the process) enable distributed program planning and control across multiple interrelated projects, as long as all of the projects in the program use the same DCM system. Distributed program planning and control require that the enterprise planning and control system manage the interrelationships of schedule-items across projects, not just within a project. This is the essence of distributed program management for this dimension of project planning and control.

Distributed planning and control are most critical when a complex project is organized into subprojects while at the same time the total project, the subprojects included, is managed as a program. This was the case for the Fast-Food Robot project, as illustrated in Figure 14–2. One of the subprojects, Arms and Fixtures, is defined as a step in the primary project, the Fast-Food Robot, and then managed as a subproject. The Arms and Fixtures project has its own team and project manager, who focus exclusively on the development of the arms and related fixtures for the robot. The project's objective is to create a working prototype for the arms and fixtures, based on the requirements specified by the Fast-Food Robot team.

> In distributed program management, interdependencies among projects are managed by linking the dependencies of schedule items across projects.

The Arms and Fixtures project consists of four major steps in Phase 2: Detailed Mechanical Design, Part Fabrication, Prototype Build, and Prototype Test and Revision. Just like any other enterprise project, there are interrelationships among the schedule-items within the Arms and Fixtures project, but there are also several critical interrelationships between this project and the Fast-Food Robot project. One of these is illustrated here. The Part Design task in the Arms and

F I G U R E   14–2

Illustration of Arms and Fixtures Subproject within the
Fast-Food Robot Project at CRI

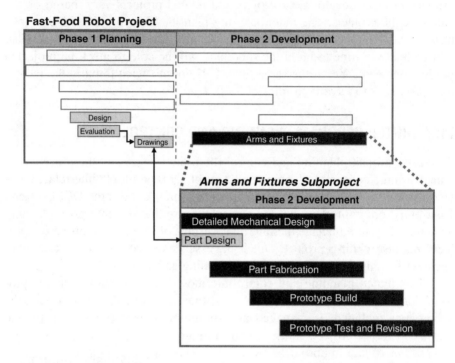

Fixtures sub-project depends on completion of the Drawings task in Phase 1 of
the Fast-Food Robot project. In distributed program management, interdependencies among projects are managed by linking the dependencies of schedule-items across projects.

Cross-project interdependencies are also important in synchronizing project planning and control from one version of a product to another, although these are not as deep as they are for subprojects within a complex program. The same is the case for technology projects integrated within product development projects in a broader program.

DISTRIBUTED PROGRAM TEAM COORDINATION

Chapter 11 introduced the concept of networked teams, and then the concept of networks of networked teams. If you don't recall these concepts, you'll find it helpful to go back and review them, since we're about to build on both concepts. In distributed program management, the project teams from all the projects in the program coordinate in two primary ways: common documents are shared across all projects, and project team members from all projects self-organize to collaborate on common tasks.

Let's look first at program-level collaboration on shared documents. Multiple teams in a program frequently need to use the same documents—specifications, drawings, plans, etc.—and it's critical that they all have access to current and accurate versions of these documents. Providing this common access to accurate and up-to-date documents has historically been difficult, especially in complex and fast-paced programs. With distributed program management as part of development chain management, there is a common enterprise system for collaboration based on common access to common project content, not just within a networked project team but across all teams in the program as well. In fact, access to critical project documents across the program is just as easy as it is within a project.

Figure 14–3 illustrates how critical documents are shared across projects within a program at CRI. The mechanical arm specifications and design documents are important to several projects within the Retail Robot Program. These shared documents are linked to the relevant steps in each of the projects and can be accessed from these steps by those working on them. In the Fast-Food Robot projects I and II, the mechanical arm step is linked to these design documents, while in the Arms and Fixtures subproject they are linked to the detailed mechanical design step for the arm.

Two other projects in the Retail Robot program also use these mechanical arm specifications and design documents. The Advanced Mechanical-Functions Technology project is actively involved in the design and test of the new arm, and the members of that project access the mechanical arm documents from the design validation step. Likewise, the Robot Factory project personnel reference these critical design documents in the process design step of their project.

Al Abrams, the project manager for the Robot Factory project, commented on the new approach to distributed program management. "This is far superior to the way we went about it previously. Usually, manufacturing was excluded from the product design, and we frequently worked on design information that we found out later had been revised, so we had to redo a lot of work. Yet, when we were included in the project team, we wasted a lot of time on issues we were not concerned with, just so we could get the information we needed. I'm not sure which was worse, but I do know that distributed program management is much better. We can focus on our project—building

F I G U R E   14–3

Distributed Program Collaboration with Shared Documents at CRI

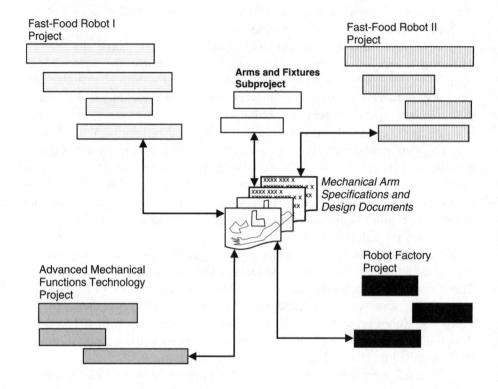

a new factory—but we have seamless access to current specs whenever we need them."

The other way in which project teams collaborate across projects in a program is by self-organizing around specific project work elements. In Chapter 11, we discussed the concept of groups of team members self-organizing within a project. Now we'll apply this concept across projects within a program.

This self-organization across projects is best illustrated by the example in Figure 14–4. Team members across all teams in the Retail Robot Program who are working on the mechanical arm for the Fast-Food Robot are able to self-organize and work together across the boundaries of their specific projects. They can collaborate on the specifications and design documents just as though they were a team specifically focused on this collaboration. They can also collaborate by working together on open issues and action-items for the

Distributed Program Team Collaboration at CRI

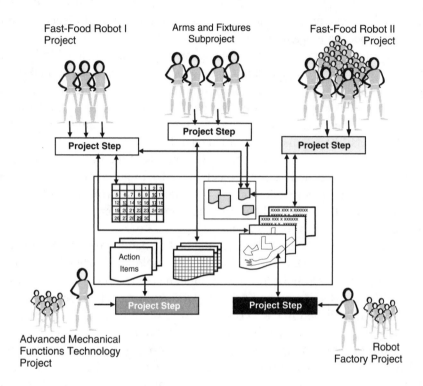

design of the robot arm, posting bulletin-board notices regarding the arm and maintaining calendar-items for their shared work.

All involved in the Retail Robot Program use the virtual workspace instead of a physical workspace to coordinate their efforts. As illustrated in Figure 14–4, they use the project step in each of their own projects as the communications mechanism that links them to the common documents, calendar-items, action-items, issues, bulletin boards, and other mechanisms for organizing and communicating.

INTEGRATED PROGRAM FINANCIAL MANAGEMENT

Integrated program financial management is the third dimension of distributed program management. This dimension entails the integration of the new product financial plans and project budgets for all projects included in the program.

In a distributed program, the financial information pertaining to the various program elements cannot be managed independently, since the projects in the program are related. Such independent management of financials would likely result in revenue being double-counted from multiple projects, and costs being improperly allocated among the related projects.

Integrated program financial management uses the financial information created in individual projects and consolidates it across projects, making the appropriate adjustments and additions to complete the financial view of the program. In a development chain management system, this consolidation can be achieved by building financial models for programs that are based on the financial models that we discussed in Chapter 13.

When the program includes multiple versions of the same product, financial integration is needed across projects in order to get a valid financial projection of revenue from each version. This integrated projection of revenues needs to address the issue of how revenue from subsequent versions will cannibalize revenue from prior versions.

When the program includes multiple and disparate projects that make up the whole, as in our Retail Robot Program example, the costs of each project need to be summarized in order to determine the overall program cost. In summarizing these costs, it may be necessary to make some adjustments according to standards established in each company for program financial analysis.

Probably the best way to visualize integrated program financial management is through the summary financial projection example shown in Figure 14–5 for the Retail Robot Program. The example shows an eight-year view of the program: three years leading up to the first revenue year, which CRI refers to as Year 1, and then the first five revenue years. This particular program summary includes the revenue projections, program costs, and other projected costs in order to determine the estimated profit for the program as a whole.

Program revenue comes from five products in the two product lines. The Fast-Food Robot I is expected to be the first product to market in Year 1, with $8 million in revenue. These projections are the same as we saw in the Fast-Food Robot revenue projection in Figure 13–3. The Fast-Food Robot II is expected to be released in Year 3, and to cannibalize revenue from the Fast-Food Robot I, since it will have more advanced features. Part of the program plan is to phase out the Fast-Food Robot I in year 4. Similarly, the Fast-Food Robot III is planned for release in Year 5. Note that the total revenue from all versions of the Fast-Food Robot product is the same as the projection presented in Figure 13–3, since that projection was really a consolidated estimate. Otherwise, the total revenue projection would have underestimated the total opportunity, and the Fast-Food Robot I might not have

Integrated Financial Projections for the Retail Robot Program at CRI

				Program Year				
	-2	-1	0	1	2	3	4	5
Revenue								
Fast-Food Robot I				$8,000	$44,500	$99,360	$158,025	$0
Fast-Food Robot II						$66,240	$368,725	$565,400
Fast-Food Robot III								$141,350
Retail Robot I					$12,000	$40,000	$50,000	$20,000
Retail Robot II							$30,000	$90,000
Total Revenue				$8,000	$56,500	$205,600	$606,750	$816,750
Cost of Sales				$3,722	$24,295	$86,352	$260,903	$367,538
Program Costs								
Fast-Food Robot I	$692	$4,898	$4,072					
Fast-Food Robot II				$1,000	$1,000	$2,500		
Fast-Food Robot III						$500	$3,500	$4,000
Retail Robot I	$500	$2,300	$3,500					
Retail Robot II				$750	$1,200	$1,900	$750	
Voice-Processing Tech.	$2,000	$2,000	$1,500					
Mechanical Systems	$750	$1,850	$1,350					
Other Technology	$300	$300	$300					
Robot Factory		$250	$400					
Fast-Food Sales Channel		$350	$475					
Sales & Marketing				$3,200	$22,600	$82,240	$242,700	$326,700
Administration				$800	$5,650	$20,560	$60,675	$81,675
Net Profit	-$4,242	-$11,948	-$11,597	-$1,472	$1,755	$11,548	$38,223	$36,838
Cumulative Profit	-$4,242	-$16,191	-$27,787	-$29,259	-$27,504	-$15,956	$22,267	$59,104

been developed. Revenue for the two versions of the Retail Robot is included in the program revenue projection starting in Program Year 2.

Program costs for all projects in the Retail Robot program are summarized in the next major section of the example. The Fast-Food Robot I costs are the same as were estimated in Figure 13–7, except for the allocated technology costs, which are now broken out separately in the program summary. The Fast-Food Robot II development cost begins in Year 1. Note that when we first looked at the Fast-Food Robot development costs in Figure 13–5, they were estimated as continuing R&D costs. In the program view, these costs are now estimated for the Fast-Food Robot II and Fast-Food Robot III development projects. Program costs for the Retail Robot program also include the development of Retail Robots I and II.

Technology development costs that are part of this program are also included. These are the costs of Voice-Processing Technology, Mechanical Systems, and Other Technology. These development costs were previously allocated to the Fast-Food Robot project, but they are broken out here in the program view.

Program costs also include the costs of related projects, specifically the new robot factory and the new fast-food sales channel. CRI's practice for program costing is to include only the incremental costs for related programs, not the capitalized costs or overhead normally incurred.

In total, the Retail Robot program will require a cumulative investment of approximately $29 million through Year 1 before it begins to be profitable. After the first five years of revenue, the program is expected to recover this investment and make an additional $59 million in cumulative profit.

REQUIREMENTS

Distributed program management requires that a DCM system be implemented across all projects in a program in order to provide the capabilities needed. Enterprise project planning and control integrate the project plans for all projects in the program. Distributed project plans cannot be integrated if all projects use their own separate tools for planning. Networked project teams and the project networking system must be implemented in order to create a network of networked teams whose members share common documents and self-organize around common work. Finally, the DCM system must be capable of integrated financial planning and budgeting for new products in order to integrate financial information across all projects in a program.

> There needs to be a clear understanding throughout the organization of the difference between a project and a program. This is not always clear.

When distributed program management encompasses all related projects, including some that may be external to product development per se, then the DCM system needs to be implemented outside of product development as well. There are some significant advantages to doing this, and some companies are implementing a common DCM system for all project-based management, such as IT projects, capital projects, and internal improvement projects. All projects benefit from the same improvements as product development and can now be integrated as related projects in a common program.

Organizationally, distributed program management requires a clear distinction between the roles of project managers, subproject managers, and program managers in order to avoid confusion. There also needs to be a clear understanding of the differences in how projects and programs are managed and approved.

Finally, distributed program management is a management process. There needs to be a clear understanding throughout the organization of the difference between a project and a program. This is not always clear. Clarifying this difference requires that the process definition of each be sufficiently defined, and that guidelines be created to help development managers decide when it's appropriate to create a program instead of a project in order to get sufficient consistency across the company.

BENEFITS

Distributed program management enables a set of related projects to be managed as a whole, dramatically improving execution and results, particularly in complex programs.

1. Program managers can manage progress of subprojects as part of a top-down program plan, ensuring that all subprojects are synchronized.

2. Program managers can distribute the responsibility for specific program elements, steps, and tasks to sub-teams. Specific accountabilities lead to a clearer focus on the individual aspects of the project.

3. Members of teams working on related projects can share common information and documents, minimizing the chance for errors and mistakes and reducing delays.

4. Members of geographically separated development teams working on related projects can self-organize "virtually" in order to coordinate their work.

5. Program-level financial planning and budgeting is automatic with integrated financial management, eliminating the need for manual efforts to consolidate information across projects, and thus the chance for errors.

SUMMARY

Distributed program management is the ability to distribute the management of related projects while simultaneously managing the projects as an integrated program. We looked at different types of related projects that are included in product development programs, including multiple versions of a product, complex projects with subprojects, technology development integrated with product development, and related process projects. There are three dimensions to distributed program management: distributed program planning and control, distributed program team coordination, and integrated program financial management. We looked at each of these and illustrated them with the Retail Robot Program example.

Distributed program management can resolve one of the longstanding dilemmas of major, multidimensional development projects: how to maintain a simultaneous focus on the parts and on the whole. In technology-intensive development, details are profoundly important, and the detailed elements of a new product, technology, or production process are best executed by tightly focused, detail-minded teams. At the same time, however, major development projects are extremely demanding exercises in coordination, in fitting together all the highly refined pieces of the puzzle. Distributed program management supports the effective networking of networked project teams, self-organizing around both project details and project interrelationships.

15

Collaborative Development Management

Collaborative development—developing products across corporate boundaries by collaborating with a range of external partners—is a recent product development trend and a key element of the new R&D Productivity Generation of development management. Companies realize that they do not have the internal resources and competencies needed to do everything they want to do, and they are increasingly turning to development partners such as customers, suppliers, contractors, and codevelopers.

Collaborative development will eventually change the very nature of product development, which will become less constrained by the availability and capabilities of internal resources, as external resources will provide the ultimate flexibility. Product development investment will become more variable, instead of the predominantly fixed cost it is today in most companies. However, there is a lot to be done in order to get there. Companies need to wrestle with issues such as increasing their outsourcing of R&D and understanding the true cost of internal development. They also must put in place the infrastructure necessary to manage collaborative development. This is the primary focus in this chapter.

Prior to this new generation of development management, and the advent of DCM systems with partnering capabilities, collaborative development was an elusive goal because the necessary infrastructure to manage collaboration was not yet in place. Companies wanted to collaborate seamlessly with outside partners, but the reality was that they could not even achieve seamless internal collaboration because they lacked the common systems and processes across their own enterprise. These missing systems and processes are now within reach, and they will allow companies to build on the capabilities

> Collaborative development will eventually change the very nature of product development, which will become less constrained by the availability and capabilities of internal resources.

we discussed previously, in particular two important new capabilities of the next generation: enterprise project planning and control systems and networked project teams. These two capabilities enable project teams to coordinate their project schedules and to work collaboratively. In the last chapter, we looked at how these capabilities can be extended across multiple related projects to an entire program within a company. We'll now take these coordinative capabilities one step further, and extend them to outside partners that participate in developing new products.

The primary focus of this chapter is on how new systems and processes create the infrastructure that enables collaborative development management. In addition to infrastructure, companies also need a collaborative development strategy, but the strategic and policy issues in collaborative development are beyond the scope of this book. We'll look at the alternative systems for collaboration, and the collaborative services they provide, but first let's understand the various categories of collaboration.

CATEGORIES OF COLLABORATIVE DEVELOPMENT

It's helpful to categorize the different types of partners for collaborative development, since the collaboration required is different for each. We'll review each of these partner types, using the Fast-Food Robot project at our hypothetical company, Commercial Robotics, for purposes of illustration. The Fast-Food Robot was the first project at CRI to use collaboration extensively, with 10 partners helping to develop the robot. We'll discuss the different types of collaborative development systems later in this chapter, but for the sake of our example, CRI uses a partnering extension to its development chain management (DCM) system, allowing the team to designate specific items to be shared with specific collaboration partners.

Collaborative Development for the Fast-Food Robot Project at CRI

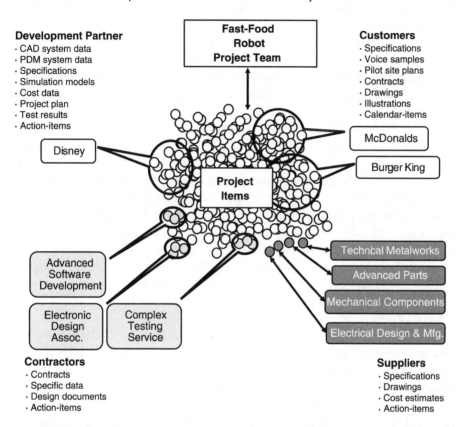

The Fast-Food Robot team collaboration is illustrated in Figure 15–1. The product is complex, and the project team manages more than 5,000 project items, including documents, drawings, financial data, schedule-items, action-items, and calendar-items.

Customers

Collaborative development gets customers much more involved in the development of new products, both by soliciting their insights and by gaining their early support for new products. There are many ways in which a company's customers can be drawn into participating in the development of a new product. At the high end of involvement, a few lead customers can be recruited to provide deep collaboration on a new product, where they spend significant time helping to guide the new product from concept through requirements and into final design and possibly even testing. With this approach there is likely to be a lot of information shared.

At the other end of the customer-involvement spectrum, new product teams could solicit customer feedback on current and prospective products, then apply the feedback in updating product features on a regular basis. These customer forums can either be moderated (edited), or not, depending on the project team's objectives.

The Fast-Food Robot team at CRI collaborates closely with its two lead customers, McDonald's and Burger King, although the team goes to great lengths to avoid sharing its information about one company with the other. The team shares the current version of the robot's product specifications with each customer, but maintains an ongoing commentary forum on these specifications with each customer separately. In addition, they share product drawings and illustrations with both customers, but since each Fast-Food Robot is customized for each customer, the two companies see somewhat different versions of the drawing and specifications. The team also maintains blueprints and specifications on a generic version of the Fast-Food Robot from which both the McDonald's and Burger King variations are derived, but it doesn't share this generic version with either customer.

One major area of collaboration is the development of the robot's most advanced technology: the voice-processing module. There is a wide range of spoken words, accents, slang, tone, and volume that needs to be accommodated. The Fast-Food Robot team enlisted the collaboration of its two lead customers to solve this problem. Each lead customer is collecting several thousand voice samples from actual customers at locations throughout the U.S. (CRI decided to wait until later to deal with customer voice processing in foreign languages.) The two lead customers posted signs in selected stores notifying customers that voice recordings were being made as part of a campaign to improve service, and gave customers the option of not being recorded.

These digital recordings are organized by each lead customer and posted daily to the project's DCM partnering system. The Fast-Food Robot project's voice-processing technology team uses these recordings daily for voice recognition testing. Donna Trotman, the leader of the voice-processing technology team, commented on the value of such a large amount of voice data: "We're able to test 50 to 100 voice samples each day of actual customers placing orders, asking questions, or just conversing. The variation in these samples is amazing. Initially, we were only able to get one in five of the voice samples correct, but now we're closer to four out of five. By collaborating online with our lead customers, we're able to process these recordings without delay, sometimes on the same day they are made. This helps us react immediately to what we learn, instead of doing the samples in batches. For example, we learned very quickly that we needed to get the real order information in addition to the voice sample, so we added that requirement to the protocol by the third day. I expect that we cut our development cycle by at least six months, possibly more, through this collaboration."

The Fast-Food Robot team also collaborated closely with its two lead customers on the customized aspects of their robots. One of the primary differences in the robots was how they planned to handle voice-processing problems. The McDonald's robots will all have a touchpad order-selection box on the counter that is integrated with the robot. When the robot cannot understand the customer, it asks the customer to select an item and quantity using the touchpad.

Burger King, on the other hand, wanted to create as much of a "human experience as possible," and had its robots customized to default to human intervention when they cannot recognize what the customer says. A robot that cannot process a spoken order will default to a Burger King customer-support center based in Maine, where the request will be instantaneously replayed for a support staffer who will converse directly, but remotely, with the customer, as if it were the robot talking. The Burger King model even alters the human voice to sound like the robot's voice.

The Fast-Food Robot team designed variations of the robots to fit each of these customer configurations, and used the partnering capabilities of its DCM system to share these designs and all related specifications. They used the same design tools as their "partners" at McDonald's and Burger King. Again, the Fast-Food Robot team was very careful not to share information from one company with the other. The general design information for the rest of the robot was common across the Fast-Food Robot team and both lead customers. To the team, of course, the design information was all transparent, and managed by the DCM system.

The Fast-Food Robot team also worked individually with each lead customer to design the robot heads. "McDonald's didn't want customers to think that 'Charlie' (their robot) had just taken a job down the street to work for Burger King," joked Lynda Stevens, who managed the mechanical design for the project. So each customer selected from a set of customized heads and paid for their development.

Lynda explained how this worked: "The head designs are actually done by our external development partner, who is also a collaborator on the Fast-Food Robot team. So the lead customer, our team members, and the external designer all collaborate by using the same designs and information. Our

DCM partnering system automatically manages the information for the right customer. The only thing we had to insist on was that everyone used the same CAD and simulation tools."

Development Partners

A project team may decide to partner with another company to develop a product. In this case, the external partner has broad responsibility for a portion of the product, and typically that portion is managed as a subproject within the overall project, similar to what we saw in distributed program management, except that this is an external project. Typically, the external partner is under contract and is paid for its services, but, in the future, external partners may have some or all of their earnings based on a percentage of the product's revenue.

CRI works closely with Walt Disney's Animatronics Division, which is Disney's contract research and development group for commercial products outside of Disney. This group has exceptional skills in making robots look and act like humans. CRI contracted with Disney for $1.2 million to design the head and upper body of the robot, including all of the subtle movements and the head design alternatives. The Fast-Food Robot team prepared the initial specification, and after the first draft the Disney designers participated in all subsequent revisions. The Disney team prepared the first designs for the head and upper body, and shared them with the Fast-Food Robot team as well as the two lead customers.

In addition to document and file collaboration, the developers from CRI and Disney use other collaboration capabilities quite extensively. They regularly use a common calendar, which is actually a portion of the overall Fast-Food Robot project calendar used by the CRI developers, but looks like a special calendar to the Disney developers. The Disney and CRI developers also post items to the bulletin board and use the DCM system's action-item generating and tracking capabilities. Halfway through the project, CRI's collaboration with Disney Animatronics included more than 100 documents or versions, 50 designs and drawings, 20 simulations, dozens of calendar-items, and hundreds of joint action-items. These were all integrated into the overall project information for the CRI developers, so they didn't need to visit a separate web site for each partner.

CRI had previously established a set of policies for collaborative development, and these were applied to the Fast-Food Robot project. CRI enables its two lead customers, McDonald's and Burger King, to collaborate with its primary development partner, Disney, through CRI, but prohibits them from collaborating directly with Disney on robotics for a period of five years. CRI foresaw that while allowing multiple collaboration partners to work as part of the same team would be efficient, there was a risk that CRI could get cut out as a partner at a future date.

With this agreement in place, CRI lets its lead customers work directly with Disney to develop the custom heads for the Fast-Food Robot. CRI plans to develop up to three custom heads for each customer, depending on volume, and the design of any additional heads will be charged as a design special. Once McDonald's and Burger King got going with the robot design, they each started requesting more heads. So far, Disney has received an additional $450,000 from McDonald's and $350,000 from Burger King for these specials, with CRI taking a 20 percent markup on them.

Contractors

Using outside contractors is the traditional form of external development, with contractors used like temporary workers who work on site as part of the project team. With systems-enabled collaborative development, however, outside contractors will be better integrated into projects and will not be constrained by distance, since they will be able to work remotely from the project team.

The Fast-Food Robot team uses its DCM partnering capabilities to collaborate with three contractors. Advanced Software Development, based in India, does some of the software development work that CRI couldn't staff internally. The contractor is required to use CRI's DCM partnering system as its primary development environment, so that the Fast-Food Robot team has immediate access to all the software developed each day and is able to communicate with the contractor using various collaboration capabilities.

Electronic Design Associates, based in Scotland, provides consulting services on the design of the robot's main processing unit. It has access to all electronic designs and uses the same CAE tools that Commercial Robotics uses, making the collaboration easier. Complex Testing Services, an external partner, will put a prototype robot through various environmental tests in order to establish its environmental tolerances. At this point in the project, Complex Testing Services is just evaluating the specifications and preparing a test plan. They are also communicating directly with the two lead customers to understand expected environmental conditions, although they

> With systems-enabled collaborative development, outside contractors will be better integrated into projects and will not be constrained by distance, since they will be able to work remotely from the project team.

are doing this blindly. They don't know the names of the lead customers, or which has what requirements.

Suppliers

The integration of suppliers into the development process is a natural extension of collaborative development. Suppliers can take on some of the burden of the product's design, enabling the OEM or design company to complete more products. In many cases, suppliers also have a lot to offer in terms of improving the design of a product, since they better understand the technology of specific components or materials. There are several unique issues that need to be addressed with supplier partners. When a supplier collaborates on the development of a product by providing some component, it is usually designated as the sole supplier of that component, eliminating the opportunity for Purchasing to reduce costs through strategic sourcing. There is also the issue of whether the supplier gets paid for development work or includes this cost in the future price of the components it provides.

The Fast-Food Robot team also has four suppliers collaborating on the development of the robot. CRI makes a major distinction between development partners or contractors and supplier partners. The first two are generally paid for their services, while supplier partners are generally not paid for development but are guaranteed the right to provide some portion of the supply of the component for the product. Supplier partners' design services are part of their total product offering to their customers.

Chris Taylor is CRI's vice president of R&D, and supplier partners are part of his collaborative development strategy. "This is a way for us to capitalize R&D costs instead of expensing them. We only pay supplier partners for their design services when we purchase the components from them. They share the benefit and risk in the success of our product with us. We have more new product opportunities than we can afford to fund because we are limited in how much R&D we can charge to our income statement. But with supplier partners the development cost is theirs, not ours, so it doesn't go on our income statement. On the Fast-Food Robot project we're saving approximately $1 million working with supplier partners, which is more than 10 percent of the project budget prior to allocated development costs. If we can do this across all of our projects, we can increase our R&D investment by more than $10 million without incurring any additional cost."

On the Fast-Food Robot, a company called Technical Metalworks is designing and will manufacture the robot body and various other metal casework. Another supplier, Advanced Parts, is designing and will manufacture all of the movable components for the robot. Still another supplier, Mechanical Components, is designing and will manufacture some of the simpler mechanical components. These suppliers are supervised by Lynda Stevens and her subteam, who use the self-organizing capability of the DCM system to work with each of these three supplier partners. As part of their partnership with CRI, the three suppliers have agreed to use the same mechanical design tools, and every design revision is posted immediately on the DCM system. Lynda also requires each of the supplier partners to keep its expected component-cost estimates current with every design revision, and she uses these new estimates to automatically update the product cost.

Electrical Design & Manufacturing is the sole contractor for producing all of the electronics for the Fast-Food Robot, and is doing all of the electronics-related design work. The company posts the updated electrical designs on the Fast-Food Robot DCM system every time there are changes. Electrical Design & Manufacturing is required to use CRI-approved components. The use of any other components requires approval from CRI's electrical component engineer and electronics purchasing specialist, and they all use the DCM partnering capabilities to manage this process.

Channel Partners

Sales-channel or marketing partners can also benefit from collaboration with the product development team, especially toward the end of development, but also just prior to a product's release. This collaboration can help sales and marketing partners prepare to sell the product, participate in release planning, understand the functions and features of the product, and so on. Some companies find this collaboration helpful, especially where they have a very close relationship with these types of partners, but others are more cautious. They want to avoid scaring their channel partners with the problems that are typically associated with new product development, and prefer just to give them periodic updates emphasizing the positive. CRI is selling the Fast-Food Robot directly, and is not using any channel partners.

COLLABORATION SYSTEMS SERVICES

There are a range of system capabilities necessary to support a collaborative development infrastructure, and we'll examine these by the services they provide. Figure 15–2 provides a framework that illustrates the different types of collaboration systems as layers of collaboration services. Each higher layer generally provides the services of the layers below, plus the new services de-

Collaborative Development Systems Services

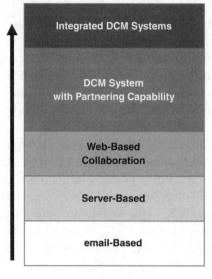

• System-to-system integration

• Schedule coordination
• Automatic integration to main project
• Many to one partner management
• Automatic documents and file sharing
• Self-organizing teams

• Partner interaction
• Document and notice posting
• Team collaboration

• Document and file sharing

• Communications
• Document and file exchange

scribed. While this illustration is a bit oversimplified, since all collaborative systems are different, it does provide a useful framework for classifying and understanding the differences.

Collaboration Based on Email

The base level of collaboration services is provided by email. This enables members of an internal project team to communicate with external collaboration partners and exchange relevant documents and files. Without this basic email service, collaboration would not be practical, and it is the reason that collaboration has grown in recent years.

Email services, however, are subject to the defects and deficiencies we discussed in Chapter 11 regarding distributed communications within a project team. There is little or no structure to email communications, and this lack of structure typically leads to costly and time-wasting overcommunication. At the same time, it often results in someone in a "need to know" position failing to receive a vital communication. With distributed communications via email, there is no content control, and when used for external communications there can be security problems. With distributed communications through email, anyone on the internal team can send any document to external partners, based on his individual judgment on what is appropriate, without any approval or filtering. Generally speaking, collaboration starts with email, and it continues along until it gets to be frustrating or someone makes an embarrassing mistake.

Using a Shared File Server

With the next level of services, server-based collaboration, a jointly accessible file server is set up for each partner, and the internal team files documents to be shared on this server. This addresses the inefficiency and security concerns of email document management, but also poses some significant inefficiency of its own. Individual file servers generally need to be set up for each partner, and documents must be posted to each server in addition to the internal server used by the project team. In the Fast-Food Robot team example, 10 servers would be needed. The internal project team would have to work with these redundant servers, regularly checking to see what is new and posting new documents and updates to all the relevant file servers.

File servers generally don't organize information, except to use folder structures and folder titles to file and locate documents. While this may work with a limited number of documents, it is generally not scalable as the number of documents increases. A shared file server also poses the same risks of content control and security presented by email. Some companies will move to this level of services from email, but others will skip this step and go directly to a higher level of collaboration services once they determine that collaborative development is important in their future.

Collaborating with a Web-Based Workspace

Web-based collaborative workspaces provide a web site for collaborating with a partner, adding a number of additional capabilities. Documents and files can be stored directly in this workspace, which generally has a more organized access than a file server and sometimes provides version tracking. Collaboration tools such as real-time meetings, threaded discussions, and bulletin-board postings are sometimes provided for the collaboration workspace. In some cases, there may also be capabilities for providing alerts to users of the web site, and some web-based collaboration workspaces also provide project planning and scheduling tools, enabling the scheduling and control of joint project tasks.

A web-based collaborative workspace provides a higher level of collaboration services, greatly enhancing the abilities of partners to work together. In the example illustrated in Figure 15–3, the project team established four collaboration workspaces, one for each partner. Members of the internal team can then go to the appropriate workspace in order to share documents, post no-

F I G U R E 15–3

Collaboration Workspaces with Four Partners

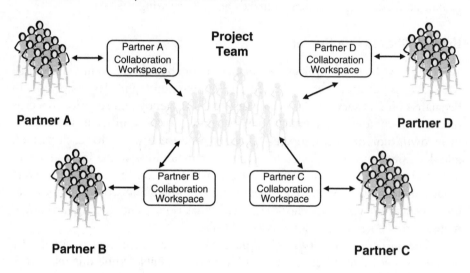

tices, and communicate project work. Partners use the Internet to access the partner workspace and do the same.

The principal drawback to this approach is that these workspaces are not fully integrated into the DCM system and project networking for the internal project team, so they need to access five separate systems—their own internal system and each of the four partner sites. In the CRI example, the team would need to establish 10 partner workspaces, and many team members would need to check several of these, and possibly all of them, daily. The team also needs to post the same documents and information to multiple web sites if those documents and information are common across multiple partners.

In other words, this approach may be suitable for collaboration on small projects, as well as a way to begin collaboration on a limited basis, but it is not sufficiently scalable to efficiently handle many partners on a complex project. DCM systems with integrated partnering capabilities address these scalability issues.

Integrating Partnering Capabilities into the DCM System

Development chain management systems provide an integrated platform for broad collaboration within a project team and across projects as part of a common program. With partnering collaboration, these systems are extended to external development partners. At the beginning of this section we reviewed three new capabilities that enable the DCM system to provide these collaboration capabilities.

The enterprise project planning and control capabilities of a DCM system transform project management from a technique into an enterprise-wide management process. Project planning is no longer limited to the project manager's desktop; everyone shares a common project plan. Enterprise project planning and control can be extended to external development partners as well. They can be given selected access to, and even responsibility for, specific elements of the project plan, such as steps or high-level tasks. For example, on the Fast-Food Robot project the Disney Animatronics Group was given responsibility for the head-and-shoulder design step in Phase 2, and managed it as a project within a project, as is illustrated in Figure 15–4.

This is similar to distributed program management, as described in the previous chapter. The Disney team manages its work as a separate, but related, project that is directly integrated into the Fast-Food Robot project. In this example, the drawings task in Phase 1 needs to be complete before work can be-

F I G U R E 15–4

Illustration of Collaboration on Head and Upper-Body Design within the Fast-Food Robot Project at CRI

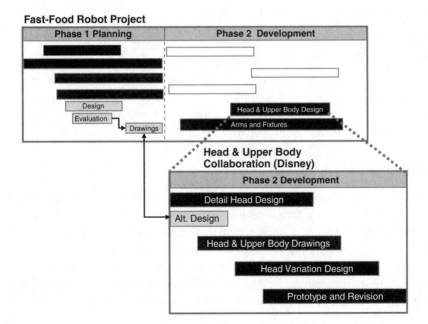

gin on the Alternative Designs task within the Detail Head Design Step. These interdependencies are managed by linking dependencies across projects, in this case an external project. The Disney team helps to prepare the project plan for its work, while considering how its plan will fit into the overall Fast-Food Robot project plan. The Disney team can then use the same project control system to report progress and exceptions that are then automatically consolidated with the other steps of the Fast-Food Robot project.

The second DCM capability that enables external collaboration is the networked project team, when extended to external partners. This enables the networked communications, project coordination, and shared content management that are illustrated in Figure 15–1. The DCM system for the Fast-Food Robot team manages thousands of items for the project as a whole, including documents, drawings, simulation models, specification documents, knowledge management documents and links, action-items, calendar-items, and even more. Some of these project items are automatically shared with the 10 development partners for the project by designating for each item or type of item which partner, if any, has access to the item in question as a shared item.

> A DCM system with a partnering capability enables the internal project team to manage and view all the project items through a single virtual workspace, unlike the previous level of services, where the team needed to go back and forth among multiple workspaces.

A DCM system with a partnering capability enables the internal project team to manage and view all the project items through a *single* virtual workspace, unlike the previous level of services, where the team needed to go back and forth among multiple workspaces. However, the external partners may need to work in multiple workspaces until they have their own DCM system that is integrated with other DCM systems.

Integrating DCM systems

The final level of collaboration services is the integration of multiple DCM systems. The primary internal project team uses its DCM system to collaborate with external partners, but instead of just being an external partner on that DCM system, the external partner has its own DCM system. The two systems are integrated, providing each partner with the ability to manage its own projects, while these projects are part of a broader external collaboration. Just as the primary project team only wants to share selected documents and information with its external partners, the same is true for the external partner: It only wants to share selected documents and information it uses on the project.

The integration of two DCM systems is illustrated in Figure 15–5. The Fast-Food Robot project team manages many items (documents, schedule-items, action-items, calendar-items, etc.) for the project, but it shares selected items (indicated by the darker color) with the Disney Animatronics project team. The Disney team also manages many items for the project that it is contracted to do for CRI, but there are also many items that the Disney team prefers to keep to itself. These items include internal correspondence, previous designs that are used as references, schedule-items that are consistent with the Fast-Food Robot project schedule but intended for the Disney developers, and the like.

Each team "sees" its own project, and the two DCM systems manage the collaboration of individual items. For example, a Disney developer may see an item from the Fast-Food Robot project on his system, and can access it, but the item really resides on the Fast-Food Robot DCM system. Similarly, the Disney team files all of its work on its own system but makes some of it available to the Fast-Food Robot project team.

The advantage of integrating two DCM systems is that it enables each company to use its own DCM system to manage its development, including resource and portfolio management, yet collaborate automatically with any number of partners. This will be especially advantageous to suppliers when development partnering is essential to their business strategy.

F I G U R E 15–5

Example of Integrated DCM Systems

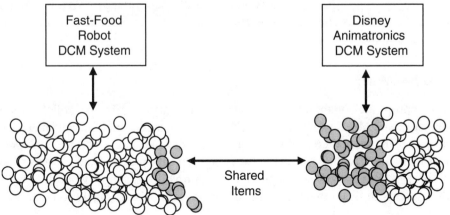

OUTSOURCING OF R&D

Collaborative development introduces the touchy subject of outsourcing R&D. Companies have outsourced many functions over the last two decades, including contract manufacturing, information technology, and customer service. Until now, however, R&D has been maintained as an internal activity, primarily performed with internal resources.

> Large numbers of companies will begin to overcome their reluctance to outsource R&D when they realize that their R&D resources have much lower utilization than they thought, and that the effective cost of their development resources is much higher than they thought.

I believe that large numbers of companies will begin to overcome their reluctance to outsource R&D when they realize that their R&D resources have much lower utilization than they previously thought, and that the effective cost of their development resources is much higher than they thought. They will see the increasing advantage of flexibility, and the increasing disadvantage of lacking some resources while having an excess of others. In addition, the flexibility of increasing and decreasing their R&D investments to fit their needs will be appealing. Finally, more and more companies will come to realize that outside partners have better R&D capabilities than they do in certain areas. Basically, these are all the same reasons that manufacturing began to be outsourced on a broad scale in the 1980s.

Each company will need to wrestle with the extent and approach to R&D outsourcing as part of its overall R&D strategy. What is certain, however, is that as outsourcing increases, so will the dependence on collaborative development management.

REQUIREMENTS

The appropriate collaboration systems are necessary for effective collaboration management. As described previously, these systems include an enterprise project planning and control system that provides interactive access to the project plan, along with a project network system that provides shared access to all project items. The basic requirement is simple: If a company does not have the systems in place to collaborate internally, then it's not likely to be able to collaborate externally.

We discussed these system requirements in terms of collaboration services, with the lower levels of services providing rudimentary support for collaboration and the higher levels providing the support necessary to make collaboration truly seamless. Companies attempting collaborative development with the rudimentary services may well encounter frustrations and disappointing results. Companies with an extensive collaboration strategy, or those moving in that direction, require a greater range of collaboration system services. When fully implemented, collaborative development management can go as far as integrated resource management, with external resource availability visible to project managers as they perform project resource planning.

Collaborative development management is more than a system; it's also a set of policies, and in the broadest sense collaborative development is a management process. Policies need to be established for qualifying and approving partners, standard partner contract terms, and expectations of partners. These policies, as well as the organizational responsibilities for approving and managing partners, become a formal collaboration process. Developing and implementing this process is essential to successful collaborative development management.

Collaborative development also changes the very economic model of product development. The investment in development becomes more variable, which has broad implications for the functional budgeting, product approval, and resource management processes. These processes need to be modified if collaborative development is to be successfully managed.

BENEFITS

Collaborative development management can have some significant strategic advantages. To review:

1. Product development is no longer restricted to the internal capabilities of a company's own R&D resources. Regular access to external partners opens up unlimited possibilities.

2. Collaborative development increases flexibility. A company can expand and contract its R&D investment as conditions and opportunities dictate.

3. Collaborative product development by supplier partners can be capitalized so as not to immediately affect a company's income; these expenses are instead spread over the term of purchase from the suppliers. This can help a company increase its product development efforts without increasing its R&D investment.

4. Greater involvement by customers in product development can increase the initial success of new products by more closely reflecting market needs.

Collaborative development management requires the appropriate systems and process infrastructure, and getting this right has some important benefits.

1. Collaboration with development partners is much more efficient. The overhead cost of collaborative development management is much lower with the appropriate collaboration infrastructure.
2. With seamless collaboration, there is very little delay or lag time with collaborative development. External partners are as integrated into the development process as internal developers.
3. The security risk of collaboration is greatly reduced. With the appropriate systems and processes, there is less risk that partners will have access to something they shouldn't.

SUMMARY

Collaborative development is a significant strategic opportunity that will eventually change the way product development is done, but there is a lot required in order to get there. In this chapter, we focused on the management of collaborative development, specifically the extensions to enterprise project planning and control and networked project teams. We reviewed several categories of collaborative development and illustrated how the Fast-Food Robot team collaborated with different types of partners. We then reviewed collaboration systems alternatives in terms of collaboration systems services, essentially a structure for determining the extent of collaboration support required.

As R&D codevelopment flourishes (and there is every reason to believe it will), a profound former constraint on what companies can achieve in terms of creating new products and technologies will be removed. But as ready access to external development resources abolishes the limitations imposed by vertically integrated internal development, a new constraint will be imposed: the ability to manage codevelopment resources strategically and efficiently. These new constraints—the ability, or inability, to coordinate the development work performed outside the walls of the enterprise with the work performed internally—will be managerial rather than material. Overcoming these constraints will crucially depend on the processes, systems, and communication paths described in this chapter.

16
CHAPTER

Context-Based
Knowledge Management

Product development depends on knowledge: knowledge about the technologies used in new products, about how to design these products, and about their markets. Isn't it ironic, then, that product development knowledge is generally not accumulated, leveraged, or managed? Every project tends to be done in isolation. The experience gained in one project is rarely applied to other projects, at least beyond the chance recollections of individual developers. Expert knowledge in specific areas may be available, but is frequently not identified or used. An experience curve is surmounted in one project, only to be squandered by the next. R&D experience and knowledge should be major assets to companies, especially those that are technology-intensive, but all too often they're not.

This is not for lack of recognizing the importance of knowledge management: R&D managers are well aware of its importance. They know that there is a big payback from better leveraging R&D knowledge and experience. The reasons for the failure to date in R&D knowledge management lie in the lack of effective knowledge management systems and processes. In particular, I believe that the primary problem is delivering the appropriate knowledge and experience to the right developer at the time that the developer needs it.

Many R&D organizations have tried to manage their knowledge by filing—or trying to file—all their relevant documents in file servers. They've used folder and subfolder naming conventions to organize the documents, and then expected that developers will go to these folders to find what they need. Unfortunately, this approach has rarely worked.

Key-word search engines offer an improvement on this technique. Instead of trying to find helpful documents by going through file servers, the developer types in search words for what he is looking for, and the search engine finds and prioritizes the documents that fit those words. The more sophisticated search engines retrieve files and documents based on searching the words within the documents themselves, instead of just key words. While an improvement over file servers, search engines are still limited. When there are large numbers of documents, they typically retrieve too many. Searching based on the word "price," for example, in order to find pricing strategies, could retrieve hundreds of irrelevant documents, and developers are not going to open and review all of them in order to find the relevant ones.

> One of the advanced capabilities of a DCM system is that it "knows" which developer can benefit from which stored knowledge, and can direct the right developer to the right knowledge at the right time. The system knows the context of the knowledge needed.

Some companies are creating internal portals to provide access to knowledge and experience. These can be very useful if put together properly, but they still don't quite solve the problem of getting the knowledge directly to the developer who needs it, when he needs it.

In this chapter, we will focus on this delivery capability, which I call context-based knowledge management. One of the advanced capabilities of a DCM system is that it "knows" which developer can benefit from which stored knowledge, and can direct the right developer to the right knowledge at the right time. In other words, the system knows the context of the knowledge needed. Before we get into this in more detail, let's review the types of experience and knowledge that are useful in R&D.

TYPES OF KNOWLEDGE AND EXPERIENCE IN R&D

There are multiple types of valuable knowledge and experience that can be transferred across projects, across developers, and across time. While these types vary widely by company, there is a general classification that can be used to describe some of the different characteristics of knowledge and experience.

Knowledge Management for Business Plans

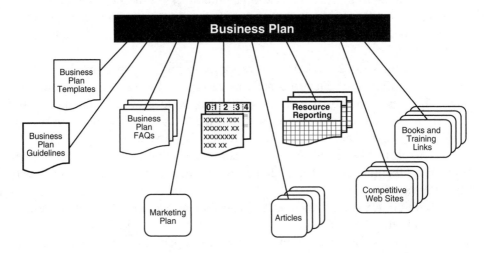

Figure 16–1 shows the different types of R&D knowledge and experience contained within our hypothetical company's (CRI's) R&D knowledge management system, such as business plan templates, sample business plans, financial models, and so on. These are illustrated as being *attached* to a specific step in a project plan, in this case the business plan step, which is why the examples illustrated are associated with a business plan. This is also the first concept of context-based knowledge management: organizing knowledge and experience around a specific project step. Knowledge related to other steps in development would be attached to those steps.

Standards, Guidelines, and Policies

Most companies have standards, guidelines, and policies to help make product development consistent with other company requirements, and also achieve some consistency across all projects. These include, for example, project-execution guidelines.

During the Time-to-Market Generation of development management, many companies achieved significant benefit from increased standardization of development by defining product development process guidelines. Instead of having every project team "reinvent the wheel," they convened groups of experts for specific project steps—for example, software design—and they defined guidelines for each step. The guidelines included recommended task plans for completing the step, cycle-time guidelines for the step, and best practices for the work to be done on that step. In some cases, guidelines were written not only for every step but also for step variations dictated by project size or complexity.

Sometimes these guidelines were too prescriptive, expecting that everyone would follow the standard exactly as written by its creator, and these usually failed. But, in most cases, the guidelines were useful. Guidelines were usually published in guideline manuals, which were later issued in electronic version. However, the use of manuals frequently proved to be ineffective. All too often, project managers and project core teams didn't want to take the time to hunt for the manuals, let alone hunt through them. They preferred to just get on with the work.

With context-based knowledge management, we are taking these guidelines to a higher level of utility, changing them from inactive reference manuals into more useful active knowledge for developers. Instead of residing in a manual (electronic or paper), the guidelines are attached electronically to the specific project step to which they relate. They can then be presented to the developers working on the specific step when they need it.

Standards, guidelines, and policies go beyond process guidelines. Companies also expect specific standards to be followed across all development projects. Examples of this include new product testing standards, software coding standards, electronic design standards, financial standards for new products, environmental standards, etc. In most companies these standards are reasonably extensive and are expected to be consistently applied, but they are not. Most developers find it frustrating to remember about or find these standards when they should be applying them. Keeping the standards current is not as challenging as distributing the current versions to the right developers and expecting them to update their working copies. Here again, there is tremendous benefit to be gained by attaching these standards to the project step to which they pertain, so that the developer will have them on hand when working on that step.

There can also be specific standards that change regularly, such as approved components and approved vendor lists. Project teams are expected to use these, but are frequently frustrated when they try to find and apply the latest version. These standards can be maintained centrally: For example, an approved vendor list would be maintained by the purchasing department, and linked to the appropriate project step, such as a mechanical design. When the developer wants to access the approved vendor list attached to that step, he gets the current version, not a copy that he previously filed based on an email with the updated list.

Here again, we also see the efficiency gains from eliminating the need for all developers to maintain the same files. It's easy for someone responsible for a standard to *publish* it by sending an email to all developers, expecting that they will read and save it, but the fact is that they won't.

In the business plan step example, CRI's business plan guidelines include a description of the elements that must be included in a business plan, analysis that needs to be done, requirements for approval of the plan, suggested responsibilities for preparing the business plan, and so on. A standard workplan and cycle time are also included in the guidelines.

Templates

Templates provide a starting point for developers when they are preparing specific documents, project plans, designs, financial models, and the like. They enable developers to start with an outline rather than a "blank piece of paper," and this saves them a lot of time. It also increases consistency across all projects and avoids cases where something was omitted by the project team simply because they didn't consider it or they weren't aware that it was necessary. All too frequently, problems occur in development projects simply because the team "didn't think about it." In this way, templates also function as a checklist of what needs to be done.

Templates may include some of the boilerplate for documents, reducing unnecessary work for developers. For example, CRI's business plan document template includes text such as "Price is expected to change by ___ percent per year, assuming little or no inflation and the expectations that _____." This boilerplate indicates to the developer working on new product pricing that the annual price change needs to be clearly defined, provides the standard inflation assumption by the finance group, and suggests adding expectations that will cause the price change. Product development is more than just filling in the blanks, however, and this can be a disadvantage in adhering to templates too strictly. It's important to use templates to the degree that is practical, and to allow developers to use their judgment in including or excluding sections of a template.

There are also a number of corporate policies that developers should comply with, and these can be distributed better through context-based knowledge management. For example, CRI has very strict policies on disclosure of confidential information and requirements for intellectual property protection. It attaches these policies to any step that they apply to, and links the entire project team to an electronic notebook where they are required to note any potential new discovery.

In the business plan example, the business plan document is the primary template. It would typically contain the outline for a business plan as section and subsection headings. Each section would include boilerplate where helpful, as well as instructions and guidelines for that section, which would be deleted as the section is completed. Most companies will find it useful to create multiple templates for different types of projects needs. For example, CRI has five business plan templates that a project manager can select from, depending on the project.

Financial Models

Financial models are a special form of template. They typically use an integrated spreadsheet containing the format for information and much of the logic for the calculations used in the model. By using standard financial models, developers are spared the work of structuring and building spreadsheets. This is something that most developers are not good at anyway, and simple mistakes in putting together these spreadsheets all too often cause problems.

An even more important advantage of using standard financial models is that the finance group can develop best-practice standard models that all project teams can apply. We explained how CRI successfully used standard financial models in Chapter 13, and won't repeat it here. In context-based knowledge management, the DCM system provides the appropriate financial model to the members of the project team when they need it.

Experience Gained from Previous Projects

Most developers rely heavily on their previous experience. They leverage documents, designs, and techniques that they used successfully in the past. They learn from their mistakes and avoid making them again. And they use the same contacts and experts to guide them that they worked with in the past. Wouldn't it be great if this experience base was shared across all of a company's developers instead of residing only with individuals? With R&D knowledge management, this *institutional* experience can be used by all developers so they can learn from the mistakes and best practices of others.

Experience is *packaged* in a variety of ways. Using samples of past work is an obvious method. CRI attaches sample business plans to its standard business plan step, but then goes even further. Each sample is annotated with the backgrounds of specific approaches, explaining why it was done that way, enabling the project team referring to the sample to understand it better. At the end of each project, the manager of knowledge management invites the project team to submit their experience, their business plans in this example, to be considered as a future best practice. The team annotates these submissions, and the best are selected based on an impartial judging committee. Those examples selected are attached to the appropriate standard steps in the knowledge management system, but more importantly CRI recognizes the team for its contribution to CRI's "Body of R&D Knowledge." The appropriate team members are given an award and stock options. CRI puts its money where its mouth is when it comes to intellectual capital.

> With R&D knowledge management, institutional experience can be used by all developers, so they can learn from the mistakes and best practices of others.

Project team members at CRI also regularly contribute to a list of frequently asked questions regarding business plans. Every time someone has a question, big or small, and finds the answer, they take an additional 5 to 10 minutes to add the question and answer to the FAQ list, so that others who might have the same question can benefit. In addition, CRI is experimenting with a business plan forum where project managers "meet" virtually to share experiences. They take turns posting and discussing various issues they're facing.

Reference Documents

Reference documents include a wide variety and type of internal and external documents; anything that could provide useful information. Typically, references can be either a document or a link to a web site containing additional information.

CRI's knowledge management system contains tens of thousands of reference documents, since everyone is allowed to post them. Each posting must be linked to at least one standard project step. In the business plan example, CRI has more than a hundred reference documents. Many of these are articles that previous teams found useful when they were formulating business plans. Some of these articles are copies of documents on the system, while others are links to articles on other internal or external web sites. CRI also includes miscellaneous documents such as memos and working notes as references.

When a body of knowledge is extensive and continually updated, it may be better to create an internal web site and link to this web site as a reference. For example, CRI's marketing department maintains a wide range of market studies, some that it did itself and others that it purchased. In addition, Marketing organizes and catalogues customer feedback and regularly conducts voice-of-the customer sessions. All of this valuable marketing information is organized into a marketing web site for internal use. The web site provides access to all the marketing information by guiding users through a marketing portal. The standard business plan step links developers directly to this marketing portal.

Project Plans and Step Planning

Project plans and step workplans provide templates for project planning, but also provide a lot of additional information. They include cycle-time estimates and resource-need estimates that are automatically integrated with other capabilities in a DCM system. We examined the use of standard step cycle times in Chapter 10, when we discussed integration with project planning standards. Standard steps are selected by the project manager when she is planning the project, and when the steps are selected, the standard cycle time is applied to the plan.

Standard steps can also include resource estimates. Resource estimates for a step are the estimated time requirements for each resource required, e.g., a business plan is expected to require 50 person-days for a project manager, 25 for the marketing manager, 50 for the financial manager, and 25 for an electrical designer. We saw this illustrated in Chapter 7, when we discussed resource requirements planning with resource requirements guidelines.

A standard step may also include a task-level workplan for the work to be done to complete that step. When selected, it provides the project team with a first cut on the tasks they need to do, who will do them, and when they need to be done. At CRI, when a standard step workplan is transferred to a specific project, it is automatically populated with specific dates based on the scheduled start date for the step and team-member names based on the responsibilities defined for team members. Developers can then modify this standard workplan to fit their needs.

A project plan template for a complete project differs from standard steps. A project template contains the suggested standard steps in the appropriate sequence for a project of that type, including the interrelationship of those steps through dependencies. When a project team leader uses a project template, the entire project plan is created automatically. She can then substitute other standard steps where necessary and change step dates.

Educational Materials

A great deal of educational material, both internal and external, is available to developers, but all too often they don't take advantage of it. They rarely want to attend training sessions until they need to apply what's being taught, and they get frustrated trying to find the training they need when they need it "right away." Just-in-time training and online training are two techniques that address these needs. The concept behind just-in-time training is to do training when the developer needs it and is ready to apply it. Online training provides training materials on the web instead of in a classroom. These two concepts complement each other, since just-in-time classroom training is rarely feasible.

Context-based knowledge management provides just-in-time training online to developers when they need it, because the system is aware of what the developer is working on and matches it to available training. It also makes various types of educational material, such as books and articles, available to developers.

> Context-based knowledge management provides just-in-time training online to developers when they need it, because the system is aware of what the developer is working on and matches it to available training.

In our business plan example at CRI, the standard step provides multiple links to useful books and training materials. For those doing a business plan for the first time, CRI has an online training program developed by an outside consultant that is linked to the standard business plan step. It also has links to dozens of educational articles on business planning for new products, categorized by topic, along with direct links to several recommended books on business planning available through Amazon.com with CRI's discount. Finally CRI's context-based knowledge management system provides a link to the online version of several books with individual links to sections.

External Information

It's sometimes easy to overlook the wealth of external information on new product development that's available on the Internet. This includes numerous technology sites with research papers, technology descriptions, and some of the latest research findings. CRI uses a number of technologies, and its lead technology experts maintain the links to the most relevant technical information.

The Internet also contains a great deal of information on competitors that can be very useful to project teams. CRI provides links to the product information of its competitors in the business plan step, encouraging developers to access this information periodically as they prepare the business plan.

DELIVERING PRODUCT DEVELOPMENT KNOWLEDGE THROUGH DCM SYSTEMS

As I stated at the outset, the primary challenge is delivering knowledge and experience to the developer when he needs it, and that is what context-based knowledge management does. Knowledge and experience are attached to the appropriate step in the project, and those who work on that step self-organize around the knowledge and information provided. When they view a step in the enterprise project plan for their project, they see all of the attached knowledge and experience grouped into the appropriate categories, because the knowledge management system is an integrated part of the DCM system. The enterprise project planning and control system provides the context for context-based knowledge management. Developers simply select the project step and the category of information they want, and they see documents and links to the relevant knowledge and experience. Let's look more closely at how this actually works, using our CRI example.

Step Libraries

How does knowledge get attached to a step in the first place? While it's great to have all of this knowledge and experience attached to each project step, it's too inefficient for each project team to create this every time for their project. To make knowledge management work for projects, DCM systems take a lesson from component libraries used in CAD and CAE systems. A component library stores a great deal of attribute information on each product component, such as its dimensions, performance characteristics, tolerances, cost, and so on. When a designer wants to use a component, he simply goes to the component library and selects the most appropriate one. The selected component, along with all of its attached attributes, is then automatically transferred into the design-in-progress. The designer now has access to all of the component's attributes and can even manipulate some attributes with the design software. For example, performance models can be created by combining the performance characteristics of the components in a design, and an estimated cost can be computed automatically based on the sum of all the component costs.

Step libraries work in a similar way. When a step is selected from a step library to be included in a project plan, all of the knowledge and experience attached to that step is automatically transferred to the project. Figure 16–2 illustrates this for a business plan step. In most cases, the link to the document is transferred, rather than the actual document. So there is still just a single incidence of each document, but it is linked to many projects. If changes are made

Steps Have Attached Information from Collective Knowledge and Experience

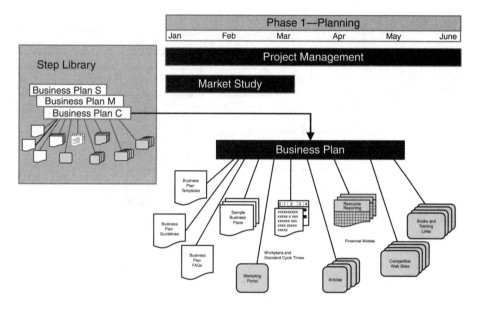

to the document, then all projects accessing it can be confident that they're getting the current version. The exception to this may be templates and models, which are intended to be modified as they're used by projects.

While it takes some time to build and maintain step libraries, just as it does for component libraries, the benefit of being able to leverage step-level project knowledge justifies the investment. Every project team will have ready access to the most directly applicable knowledge and experience related to a specific project step just when they need it most.

Linking Knowledge to Tasks and Roles

Once knowledge management is established at the step level, it may be beneficial to make the context even more specific. This can be done by attaching knowledge and experience data to high-level tasks, instead of steps. And it can be targeted even further by defining it relative to a specific development role or responsibility, which is then associated with specific developers.

Let's look at this in the context of the Fast-Food Robot project at CRI, as illustrated in Figure 16–3. The business plan step has eight high-level tasks,

F I G U R E   16–3

Tasks Can Have Attached Information from Collective Knowledge and Experience

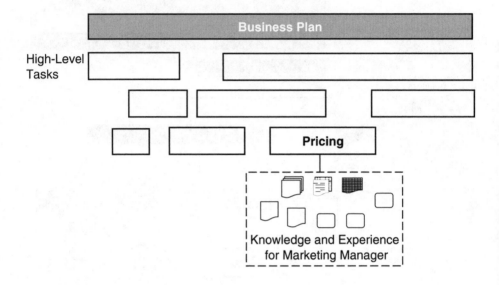

and one of them is determining the preliminary price for the new product. Attached to that high-level pricing task is CRI's accumulated knowledge and experience in preparing a new product price, and this information is further narrowed down to what will be helpful for the marketing manager, Richard Salisbury, who is working on pricing. When Richard begins working on the pricing task as planned, he sees a lot of useful knowledge and experience on pricing attached to the task and directed specifically at him.

This knowledge includes the company's standard pricing model, prepared jointly by Marketing and Finance, which has been integrated into the new product planning financial model. Richard also sees the latest guidelines for foreign pricing, as well as the current version of CRI's pricing policy. A training program for pricing is also attached, but Richard has already taken it, so he doesn't look at it. Also attached are 15 articles on pricing, as well as links to five books on pricing. Richard downloads and prints two articles that look interesting, as well as one that he customarily uses as a standard reference. It's easier for him to download and print it every time he needs it than to save it in his own files. He also sees a new book on pricing, and orders it through the Amazon.com link.

As Richard reviews some of the past pricing examples, he finds one on configuration pricing that is very creative and possibly appropriate to his project. He also reviews the FAQs on pricing and sees a couple of new items that can be helpful. Richard explains how useful all this information is: "It only took me about an hour to tap into the knowledge base and experience on pricing, and I was able to leverage quite a bit of what I found. Previously, it would have taken me a couple of days to find the information, and I wouldn't have found some of it at all. The stuff I found will save me several days of work, but more importantly, I may be able to come up with a better strategy for pricing various configurations of the robot, which could increase the revenue we get from each customer. This could be worth a lot."

COLLECTING PRODUCT DEVELOPMENT KNOWLEDGE

At present, most product development experience is lost. Developers do their work on their own desktop or laptop computers, and then neglect even to file important documents in a file server. The individual developer will have access to his own past work as he works on future projects, but others won't be able to share it. The company thus loses the opportunity to capture valuable institutional knowledge and experience. Efforts to preserve accumulated learning by requiring developers to save all critical documents at the end of a project on a file server are usually futile, and regular backup copying may preserve important documents, but won't help developers find the ones they need.

With an enterprise project planning and control system, where all the documents and all work are attached to the project, there is a simple technique for preserving important project experience. The DCM system can simply sweep the project documents and files on a regular basis and save a copy of these in some organized way, such as by project or by document type. All critical files and documents will be accessible by the entire organization.

This archiving can be extended even further by having the DCM system regularly collect all documents and save them by their use. For example, a company could sweep across all its projects and save all the pricing-analysis documents into a specific folder.

Project knowledge is different from experience, since it encompasses some judgment as to what is important and when it's important. CRI has a knowledge librarian who is responsible for maintaining the step libraries. The librarian collects documents and creates web-site links that developers find useful, and attaches these documents and links to the appropriate standard project step and high-level tasks. She also convenes expert councils every quarter who advise her on what should be included in the knowledge library.

REQUIREMENTS

In context-based knowledge management, an enterprise project planning system provides the context. All developers must have access to this enterprise system in order to access the knowledge and experience (the content), and all knowledge and experience must be attached to standard steps and tasks (the context).

In context-based knowledge management, an enterprise project planning system provides the context.

In addition, project teams need to be able to access and share knowledge and experience through a project network system. This is why these systems capabilities are considered to be prerequisites to context-based knowledge management.

To implement context-based knowledge management, a company must go though the process of creating step libraries and "stocking their shelves" with sufficient knowledge and experience so that the library will be immediately useful. From that point, they must continue to improve the library's stock of knowledge and experience as practical project experience and knowledge accumulates. Most companies will find that they already have extensive stored knowledge and experience, along with guidelines, standards, policies, models, and so on. They just need to organize and attach it all to standard steps.

With formal, context-based knowledge management, organizational responsibilities need to be clearly defined. We already referred to an administrative function such as the step librarian, but the knowledge-capturing and organizing responsibilities of project team leaders and developers also should be defined. Defining these responsibilities is part of a broader knowledge management process that defines how the process works, trains all developers in how to maintain and use the process, and measures the effectiveness of the organization's knowledge management.

BENEFITS

Leveraging knowledge and experience across all R&D development projects can yield some very significant benefits:

1. Product development teams are much more productive, and produce higher-quality work, when they learn from best practices.
2. Creating and leveraging institutional rather than individual experience keeps project teams and individual developers from repeating past mistakes.

3. R&D is fundamentally about knowledge. When knowledge, and the experience by which it is obtained, are lost or not applied, a company is allowing a valuable asset to be destroyed.

4. Context-based knowledge management gives all developers the information they need when they need it, saving them from the time-consuming task of maintaining their own reference files.

5. The efficiency of context-based knowledge management reduces time to market even further.

6. Step libraries provide the mechanism that enables the resource estimates used in resource requirements planning.

SUMMARY

R&D fundamentally depends on knowledge and experience. Most companies do a poor job of managing these vital assets, yet they recognize that there are significant benefits to doing a better job. In this chapter we looked at the challenge of delivering knowledge and experience to the developer who needs it, when he needs it, and explained why the answer lies in enterprise project planning to provide the context. We looked at the types of development knowledge, using the hypothetical Commercial Robotics Inc., as an example, and also looked at how knowledge can be delivered by step or high-level task and role.

In the conclusion to the previous chapter, we discussed how the use of external development partners promises to free companies from the restrictions posed by their internal limits: i.e., the limits of their internal competencies, the limits of their ability to invest in internal R&D. The new restrictions, as we noted, are organizational and managerial. The same is true of capturing and leveraging a technology organization's accumulated knowledge. The vital material—the recorded experience and learning—already exist. The challenge lies in capture, codification, and delivery to the right person at the right time. The repositories and delivery systems we have described in this chapter offer solutions to those challenges.

SECTION **FOUR**

**Portfolio Management
and Product Strategy**

17
CHAPTER

Stages of Portfolio Management and Product Strategy

As companies attained basic control over their product development projects during the 1980s and 1990s, they grew increasingly aware of the need to begin setting strategic priorities *across* projects. Resources were being allocated to projects on a first-come first-served basis, and managers knew instinctively that this was not the best way to do it. They also saw that they needed to better understand the characteristics of the enterprise's portfolio of projects, in order to balance risks against opportunities.

Senior product development managers began to ask probing questions about their project portfolios. What types of projects were getting resources? What was the expected *total* result of the approved projects? What was the total risk level? To begin answering these questions, they applied techniques borrowed from investment portfolio management to analyze their product portfolios. When properly applied, these techniques provided the link between product strategy and the execution of that strategy through product development. The strategic benefits of this analysis were quite significant.

Senior development managers also asked probing questions about their product development pipelines. Which were the best projects to invest in? If resources were constrained, what were the right project priorities? Should a proposed project have priority over one already approved? To answer these questions, they extended portfolio analysis to establish project priorities and then compared these priorities to some basic resource data. Here again, the benefits were significant. Product development pipeline flow was substantially improved by focusing more resources on the highest-value projects.

However, these senior product development managers soon pushed portfolio and pipeline management to its limits. They wanted more data to understand their portfolios even better, but the need to collect the data manually posed a major obstacle. They found that the information they were using to make decisions was frequently out of date because it took so long to collect. They began to see that portfolio and pipeline management should be regular and ongoing management processes, but the periodic nature of data collection and analysis prohibited this. Finally, they saw that they could make even better decisions through optimization of alternatives, and they simply didn't have the ability to keep track of multiple alternatives for every project.

Financial executives also began to weigh in on the product portfolio. They wanted to extend portfolio reporting to include more extensive financial information. They began to realize that their companies were unable to consolidate their financial projections across all projects, and that, because of this, there was a major financial disconnection between their companies' financial plans and what was actually approved in product development. In addition, they realized that there was no reconciliation between the functional budgets approved in the annual financial plan and the project budgets approved by the company's product-approval processes.

In parallel with these efforts at portfolio-level management, many companies began to put an increased emphasis on product strategy. For some, this came about when they began to formally approve new projects through a phase-based product-approval process: Executives realized that they didn't have a product strategy on which to base their approvals. With this increased emphasis on strategy came the realization that product strategy was a process that needed to be more formally integrated with other management processes. But product strategy was data-intensive, and the systems and tools to make it work were not yet available.

In my 2001 book on product strategy,[1] I introduced new concepts and techniques for product strategy and described the importance of strategic balance and portfolio management. The book described the limitations of strategic-balance and portfolio-management techniques, which in some ways were simply surrogates for what was really required. It went on to outline four envisioned changes in portfolio management and product strategy:

1. Consolidated revenue and profit forecasts will replace simple portfolio analysis.
2. Product strategy and product development will be integrated.
3. Product strategy will be balanced with development capacity.
4. The ability to simulate the outcomes of alternative decisions will enable an optimized product strategy.

These envisioned changes were not possible without new capabilities in portfolio and pipeline management, financial management, and product strategy. Fortunately, the next generation of product development systems enables these new capabilities.

Section Four of this book is about the advances in portfolio management and product strategy in the next generation—the R&D Productivity Generation—of product development, including the four envisioned changes I just mentioned. This section will explain how portfolio and pipeline management *techniques* can now be transformed into integrated, dynamic *processes* based on real-time data. We will also discuss some new concepts, such as on-demand portfolio analysis and pipeline optimization and "active" planned products.

This section will introduce a new focus on the comprehensive financial management of R&D, and will show how it integrates R&D

> Portfolio and pipeline management techniques can now be transformed into integrated, dynamic processes based on real-time data.

with financial planning and control processes. Finally, we'll look at how some of the practices introduced in my previous book on product strategy can now be automated and fully integrated with product development.

First, I'll set the stage by describing the process structure of portfolio management and product strategy, using a similar framework to the one used in the two previous sections. Then we'll look at the stages of portfolio management and product strategy, using a similar stages-of-maturity model to the one used in the two prior sections.

PROCESS LEVELS OF PORTFOLIO MANAGEMENT AND PRODUCT STRATEGY

Any complex management process can be segmented into management levels, and this applies to portfolio management and product strategy, since they're now being transformed from sets of techniques into an integrated management

F I G U R E   17–1

Process Levels of Portfolio Management and Product Strategy

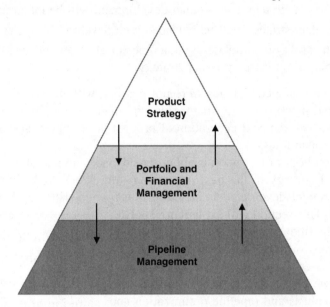

process. As was the case with resource management and project management, this process framework, illustrated in Figure 17–1, has three levels.

Product Strategy Process Level

The product strategy process is at the highest level of the framework. This is where a company creates its product strategy in the form of product line plans, platform strategies, and technology roadmaps. This top-level strategizing is now a fully integrated process in the next generation of product development, so product line plans automatically incorporate information from products already in the market, products under development, and planned products not yet in development. All of this provides a complete and integrated view of product strategy. The system supporting product strategy also manages technology roadmaps and product platform lifecycles. Product strategy simulations enable rapid examination of alternative product strategies and financial projections against development capacity.

Product strategy processes and systems incorporate formal idea management, enabling companies to track, evaluate, and prioritize product ideas from internal and external sources. In a fully integrated system, idea management can be combined with product strategy and requirements management.

Product strategy focuses on planning for new products not yet in development. These *planned products* incorporate preliminary financial projections and high-level resource estimates that are then integrated throughout the development chain management (DCM) system. This integration enables a longer-term view of the resource requirements of planned products and allows product line plans to be realistically balanced against expected development capacity.

Product strategy is performed at a higher management level than portfolio management, since its emphasis is more strategic. Typically, product line and product platform plans create annual projections of new product development costs and revenues, with yearly data over the appropriate time horizon. The objective is to plan product cost, revenue, and profit for each year of the long-term plan. An important respect in which product line and platform planning differ from the lower process levels (portfolio and financial management and pipeline management) is that the latter typically requires more detailed information, on a quarterly or monthly basis, to manage the execution of the strategy.

> Product strategy processes and systems incorporate formal idea management, enabling companies to track, evaluate, and prioritize product ideas from internal and external sources.

Portfolio and Financial Management Process Level

Portfolio management is the process of analyzing the characteristics of products under development, and sometimes those of planned products not yet in development, in order to align them with business and product strategy priorities. Portfolio management includes a wide range of techniques for characterizing projects, managing development risk, prioritizing and selecting projects, understanding what is expected from the product development portfolio, and portfolio balancing. Portfolio management provides the link between product strategy and product development. This link is sometimes referred to as strategic balancing.

As previously mentioned, portfolio management techniques were, for the most part, introduced in the later years of the Time-to-Market Generation of product development, after companies had achieved basic control over individual projects. Data were collected manually from each project, entered into spreadsheets, and then used periodically to analyze the portfolio. The next generation of product development is bringing two significant changes to portfolio management. First, integrated DCM systems automatically consolidate all portfolio information, making it available on demand, and second, they dispense with the need for manual data collection. This on-demand availability of portfolio information elevates portfolio management from a set of techniques applied periodically into a consistent, enterprise-wide management process. I refer to this transformation as dynamic portfolio management, and will explore it further in Chapter 18.

Comprehensive financial management also occurs at the middle level of the process framework I've outlined, and it is also undergoing changes in this new generation of product development. Portfolio management has long sought to take financial information into account, but, because that information was collected manually, it was very difficult to consolidate. Comprehensive financial management calls for the real-time consolidation of all financial data from individual product development projects, in order to provide a complete financial profile of R&D. The consolidated information typically includes such critical financial information as revenue projections, profit projections, product cost estimates, capital requirements, cash requirements, and R&D spending. Financial information is consolidated across varying time periods, such as year, quarter, or month, instead of the simpler totals that have been typically used in portfolio management.

> Comprehensive financial management enables the CFO to provide more proactive financial leadership in R&D—an increasingly important role.

Comprehensive financial management enables the chief financial officer to provide more proactive financial leadership in R&D, which is becoming increasingly important. In Chapter 19, we'll look at how the CFO's role in R&D is changing, the importance of a consolidated financial view of R&D, and how comprehensive financial information on R&D is aligned with planning and budgeting.

Pipeline Management Process Level

Pipeline management is generally considered part of portfolio management, but is depicted in Figure 17–1 at a lower level of process detail. Pipeline management is the process for determining priorities within a product development pipeline, based on the pipeline's capacity. It's both the strategic level of resource management and the operational level of product strategy. This is where cross-project decisions are made to balance capacity with opportunities. It includes pipeline capacity management, which was introduced in Chapter 6.

In the new R&D Productivity Generation of product development, pipeline management is extended beyond the prioritization of projects against resource capacity. The ability to simulate the results of alternative decisions about which projects, and what project variations, to feed into the development pipeline will enable even more new product revenue and profit.

In the next generation, pipeline management is highly integrated with the top level of the resource management process (project resource needs and assignments), and with the top level of the project management process (phase-based decision making).

Integration with Other Processes

We can also use the process levels depicted in Figure 17–1 to understand the interaction of the portfolio and product strategy process with other strategic processes outside of product development. This interaction is illustrated in Figure 17–2.

F I G U R E   17–2

Integration of Portfolio Management and Product Strategy Processes with Other Strategic Processes

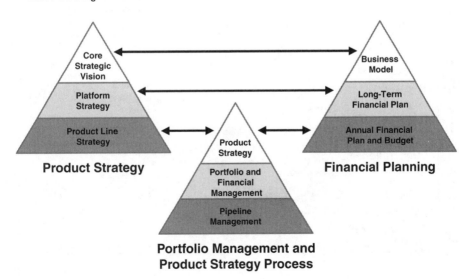

Product Strategy

Financial Planning

Portfolio Management and Product Strategy Process

The product strategy level (depicted in the center triangle) integrates with the product strategy process, and can even be considered an expanded view of the product strategy level. The product strategy level may also be integrated with the annual financial plan and budget level of the financial planning process.

Product development plans and activities are synchronized with annual financial plans and budgets in the next generation of product development, which will help companies better align critical activities and increase responsiveness to changes.

STAGES OF PORTFOLIO MANAGEMENT AND PRODUCT STRATEGY

As with other major areas of product development in the next generation, it's helpful to look at the evolution of portfolio management and product strategy in terms of progressive stages of capability. These stages are illustrated in Figure 17–3.

Companies at Stage 0 show little or no organized management of the product portfolio. Periodic portfolio and pipeline management (Stage 1) began to be exercised by companies around the late 1990s, toward the end of the TTM Generation, after sufficient improvements were made to project manage-

F I G U R E   17–3

Portfolio and Product Strategy Stages of Maturity

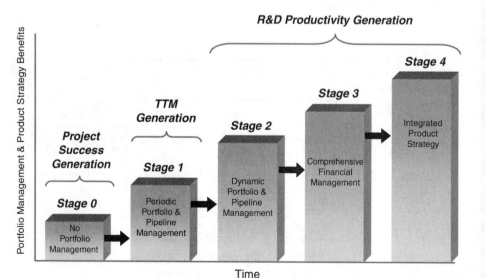

ment to support it. Because portfolio management was done periodically as needed, I refer to this stage, Stage 1, as the *periodic* portfolio and pipeline management stage.

At Stage 2, portfolio and pipeline management are transformed from a set of techniques to a management process. Dynamic portfolio and pipeline management is based on data used in individual projects, and portfolio analysis is available on-demand. This transformation is extended even further in Stage 3, with the addition of comprehensive financial management of R&D. Finally, integrated product strategy is introduced in Stage 4.

Stage 0—No Portfolio or Pipeline Management

Stage 0 corresponds to the Project Success Generation (c. 1950–1980), when the focus was on the basics of managing product development projects. Any attempts to manage projects on an interconnected basis were generally ad hoc. An individual such as the lab manager or the vice president of R&D might determine some project priorities, or a group of executives might gather to argue about priorities if contention had erupted over which projects warranted the assignment of specific people. Broadly speaking, however, new development projects were launched for no better reason than someone having what he or she thought was a good idea for a new product. So many projects were initiated that there was little visibility of the projects actually under way at any given time. In fact, when most companies moved into the TTM Generation, one of their first challenges was putting together a list of the projects everyone was working on.

The lack of a real portfolio management capability continues to plague companies even after they have adopted time-to-market-based product development practices. Project speed is not synonymous with project control. Only after companies attained sufficient control over their product development projects could they begin to move toward the first stage of portfolio management.

Stage 1—Periodic Portfolio and Pipeline Management

During the TTM Generation, a phase-review process was generally used to integrate decisions across projects, and these phase reviews prompted some very large questions. What projects fit our strategy? Do we have the right mix of projects? Have we deployed our R&D investment in an effective manner? How can we better balance our entire pipeline of projects to eliminate resource bottlenecks?

Many companies initiated basic portfolio analysis shortly after they began confronting these questions, but they tended to do it periodically. Analysts began to collect some information from all projects in order to evaluate project characteristics such as the distribution of R&D investment across the portfolio by type of project, or the return on project investment compared to project risk. An individual or a group was sometimes assigned the responsibility of being the portfolio manager. As portfolio analysis proved very helpful in guiding better product development decisions, executives and portfolio managers began to ask for more and more analysis.

The problem was that all portfolio data had to be collected manually, because product development teams were all using independent and inconsistent practices and systems to manage the information on their projects. Some team managers and members stored project data in spreadsheets on their personal computers, some created this information for presentations, and some used formal planning documents to define the information. Some important information, such as information vital to project-risk analysis, was not even collected by some project teams. This was even worse if a company wanted to manage multiple portfolios across multiple divisions.

The collection of portfolio data was time-intensive, usually taking many weeks or even months, and by the time the information was rounded up, much of it was already out of date. Projects had progressed to a new phase and had new information. Development teams revised their information based on what they were learning. And new projects were initiated. Moreover, some teams reported data to the portfolio manager that were different from the data used to get approval of their projects.

Spreadsheets were typically used as the repository of this manually collected information, and portfolio and pipeline charts were then created from these spreadsheets. In some cases, these spreadsheet tools were quite complicated, with programming that enabled data to be entered into worksheets that automatically moved the data to the right spreadsheets, which in turn produced the portfolio and pipeline charts.

Because of the need to collect all of the project data manually and enter it into the spreadsheet tools to produce charts, portfolio and pipeline management was done periodically, typically when there was some sort of problem or pronounced resource constraint. Companies managed their portfolios and pipelines on an annual basis, or sometimes more frequently, but, generally, they did not have the people available to perform this management continuously.

Data collection wasn't the only problem. Sometimes, portfolio managers were so impressed with the charts that they lost sight of the underlying data. I recall a meeting I once had with a VP of R&D. He proudly showed me charts on the distribution of R&D investment by type of project (platform, major product, minor product, or enhancement). When he finished explaining his conclusions, I asked him a few questions. Did the investment information used in the analysis include the entire investment for a product? For example, was the $10 million project with $7 million completed and $3 million remaining included as $10 million, or $3 million? Did the chart represent investment over the previous years, or the expected investment for this year? Did it include just the active projects, or past and future projects? If the chart represented the investment for this year, how much was unallocated, and where was this amount? He didn't know the answers, and began to doubt what was really represented on the charts he was using to make decisions.

Companies in this stage generally concluded that portfolio and pipeline management was essential to running the business, but they were frustrated by the labor required to collect the data, how quickly the data "spoiled," the inability to do the analysis whenever they wanted to, and the lack of flexibility in ways to look at it. These limitations formed the requirements for the next stage in portfolio and pipeline management.

Stage 2—Dynamic Portfolio and Pipeline Management

Dynamic portfolio and pipeline management is the first stage in this new generation of product development. In this stage, all the project-level data used for portfolio and pipeline management have been generated through the use of a common DCM database. The result: real-time data, on demand, for whatever analysis is needed, whenever it is needed. Portfolio and pipeline management thus become dynamic processes, enabled by an integrated system instead of techniques applied using various analytical tools.

> Dynamic portfolio and pipeline management is the first stage in the new generation of product development. The result: real-time data, on demand, for whatever analysis is needed, whenever it is needed.

At this stage, it is still feasible to have all project teams enter the necessary portfolio and pipeline data into the common system without using the full project management capabilities of an integrated DCM system. It's also possible to achieve dynamic portfolio and pipeline management to some degree at this stage, using a shared portfolio management system that does not include the enterprise project planning and control described in the previous section. But inevitably, it makes sense to have one integrated DCM system, where all portfolio data are a continuously generated by-product of project management.

Dynamic portfolio and pipeline management is more than a system. It's the transformation from a periodically used technique to a continuing management process. It's important to note, however, that it is up to the individual company to decide how it will integrate this dynamic management of the product portfolio and pipeline into the overall product development management process, and who has what specific responsibilities with regard to this integration. Each company needs to establish guidelines on which portfolio and pipeline charts will be used, and for which purposes. In addition, each company needs to train everyone who will use the charts on what they mean and how they are used. Some companies even develop a standard library of charts on the DCM system, and executives may select standard charts to track the portfolio information that is most important to them.

Dynamic pipeline management also incorporates resource information from the resource management processes. This information includes project resource needs by capacity-planning skill in addition to project assignments. Dynamic pipeline management can then be used to look at the resource loads entailed by alternative project portfolios, by including or excluding specific projects in the portfolio or even project variations. In some cases, the visibility of resource loads is extended to include planned projects (from the product strategy system and process). Resource loads can be evaluated with an eye to whatever critical resources typically constrain a development pipeline, to selected resource groups, or to all resource groups. Generally, a company needs to progress to Stage 2 of resource management in order to fully implement Stage 2 portfolio and pipeline management capabilities, but interim benefits can be achieved along the way.

Chapter 18 covers dynamic portfolio and pipeline management in detail.

Stage 3—Comprehensive Financial Management of R&D

Comprehensive financial management is considered as a separate stage of portfolio management and product strategy, because it requires much more financial data than dynamic portfolio management, and because it involves the integration of other financial processes. However, once fully implemented, portfolio management, pipeline management, and comprehensive financial management operate seamlessly.

I use the term *comprehensive* financial management of R&D for several reasons. First, financial data are more comprehensive than most portfolio information. Specifically, financial data are time-based: revenue by year or quarter, project budget by month, and so on. Financial data are also more detailed: the travel budget for marketing, for example, or tooling costs, or projected gross margin. This all makes for a lot of data. Our hypothetical company, Commercial Robotics, estimates that it manages more than 350,000 pieces of data based on 50 projects, 300 financial items tracked per project, and an average of 25 time-elements per item. When data from planned products and products already in the market are added, the volume increases to almost 1 million pieces of data. Fortunately for CRI, this massive volume of data is generated as a by-product of other activities that need to be done anyway. The source for most of the financial data is the integrated financial planning and project budgeting that we discussed in Chapter 13. Although some consolidated financial management can be done prior, it's generally best to use integrated financial planning and project budgeting to efficiently collect the needed financial information first.

As previously mentioned, one of the important advances in the new R&D Productivity Generation of product development management is the increased role of the CFO in helping to financially guide R&D. In most companies, R&D has yet to benefit from financial management and control practices as much as other functions and processes have. We'll explore this in Chapter 19, along with more discussion of comprehensive financial management of R&D.

Stage 4—Integrated Product Strategy

The final stage (at least as now envisioned) of portfolio management and product strategy includes the automation of product strategy, the formalization of product strategy as a management process, and its integration with other management processes. To this point, product strategy has been formalized to varying degrees by most companies. The focus in the next generation is on solidifying and integrating it with product development and financial strategy.

Integrated product strategy includes product platform and product line planning, whereby the roadmap of future products is drawn, but its purpose and capabilities go well beyond product roadmapping. Planned products represent active

> Integrated product strategy includes product platform and product line planning, whereby the roadmap of future products is drawn.

items which, along with their associated characteristics, are used for other purposes throughout the integrated DCM system. These characteristics include product revenue forecasts with other aggregate information used in financial planning, and high-level resource information that is used in medium-term and long-term resource capacity planning. Integrated product strategy goes beyond the mapping of planned products to include the process for collecting, evaluating, collaborating on, and prioritizing new product ideas. Idea management fosters increased creativity and an improved ability to capture and refine creative ideas.

Advanced product strategy practices build upon the information from some of the other layers of the portfolio management and product strategy framework. For example, product strategy simulation tools can use financial information from comprehensive financial management and resource information from pipeline management. Some companies may find it useful to begin to implement certain aspects of a product strategy system and process prior to adopting all the capabilities of Stage 4, and then complete and fully integrate the other aspects.

SUMMARY

Portfolio management and product strategy, taken together, are one of the three major areas of advancement in the new R&D Productivity Generation of product development. In this section, we've used a three-level framework to introduce their different process levels and how they work together. We also looked at a stages-of-maturity model for portfolio management and product strategy, describing how a company can progress from stage to stage. The objective of a stages perspective, remember, is to gauge where you are in your evolvement from lesser to more mature practices and how you compare with your competition in adoption of best practices and emerging practices.

18
CHAPTER

Dynamic Portfolio and Pipeline Management

Portfolio and pipeline management were generally introduced in the previous generation of product development after new project management practices achieved increased project predictability and heightened organizations' awareness of resource constraints. Companies recognized that they couldn't do everything they wanted to do, and had to set priorities. They also recognized that they needed to better link their portfolio of projects to their business strategy.

Portfolio management refers to the way a company selects and prioritizes a group of projects to achieve its business goals, especially maximizing the long-term value of new product investments. These investments are aimed at implementing the business strategy by achieving some balance with respect to the different types of projects, so the nature of a balanced portfolio depends on the individual business and its strategic objectives. An overall view of the portfolio is essential to a business's ability to determine whether or not its planned new products are appropriately targeted to achieve the objectives (revenue, profits, growth, market coverage, etc.) stated in its strategic plans. Project portfolios also reflect the official "plan-of-record" for a set of projects.

Pipeline management refers to the way resources are deployed and investments are prioritized to ensure that the projects chosen for investment have the resources they need to be successful. Pipeline management allows the organization to develop and understand its long-term skill-set requirements, the rate at which projects will start, and the number of projects at various stages of development. It also prevents the organization from taking on more projects than it can handle and ensures that resources are appropriately allocated between vital, medium-priority, and low-priority projects.

Section Two looked at resource management from the bottom up: How do we increase the utilization and productivity of our investment in product development? Pipeline management is from the top down: How do we best allocate resources across projects? Resource and pipeline management need to be in balance with one another.

Portfolio management aligns development investments with the business strategy, whereas pipeline management aligns resources with development investments.

Portfolio management aligns development investments with the business strategy, whereas pipeline management aligns resources with development investments. In other words, portfolio management seeks to define the new product projects to support the strategy, and pipeline management balances resources to ensure these projects get executed.

In the Time-to-Market Generation, companies got started with periodic portfolio management by manually collecting information from all projects and periodically entering this information into spreadsheets. The accumulated data were then analyzed by creating a number of charts that helped managers understand the comparative characteristics of the projects in the portfolio. This analysis was then used for periodic portfolio rationalization.

While periodic portfolio management offered important benefits, it had two major limitations. First, all of the data for portfolio and pipeline analysis had to be collected manually, which was time-consuming and inefficient. Initially, only basic data were collected, but as management began to see the power of portfolio management they began to ask for more analysis, which required more data. But because the data were frequently out of date by the time they were all collected, management ended up using "stale" information to make decisions.

The second limitation was that portfolio and pipeline management was periodic and event-driven. Analytical techniques were applied from time to time in an exercise aimed at understanding project priorities. As a result, portfolio and pipeline decisions were inconsistent and not always made at the right time.

Portfolio and pipeline management in the next generation of product development overcomes these limitations. The data used in portfolio analysis are now extracted in real time from project data, since all projects use the same database. Manual collection of data is eliminated, and all data are "up to the minute," since they're extracted when needed. Accurate and timely project data enable portfolio and pipeline management to be transformed from a periodic technique to a management process that is now an integral part of the broader product development management process in this new generation. I refer to this evolution of portfolio and pipeline management from technique to process as *dynamic portfolio and pipeline management.*

In this chapter we'll look at these processes, and how they differ from periodic portfolio and pipeline management. We'll also look at how these dynamic processes were applied at Commercial Robotics (CRI). I won't attempt a complete explanation of portfolio and pipeline management techniques, since there are some good resources available already on the subject.[1]

DYNAMIC PORTFOLIO MANAGEMENT

Dynamic portfolio management is an important transformation from periodic portfolio management done at most companies in the previous generation. It is now based on real-time integration with project data, which enables on-demand analysis. Let's look at the characteristics of dynamic portfolio management.

Real-Time Integration with Project Data

Dynamic portfolio management uses actual data from all projects, since all projects are online as part of an integrated DCM system that centrally manages the data. Project teams enter information as part of their project planning, and that information is then consolidated with similar information from other projects whenever portfolio analysis is requested from the DCM system. Other information, such as project status, is created automatically when projects are approved. Since it's probably easiest to understand this from an example, let's look at how Chris Taylor, CRI's Vice President of R&D, used this capability.

Commercial Robotics fully integrated dynamic portfolio management as part of its DCM system, and now automatically charts and analyzes data from all projects. An example of this is given in Figure 18–1, which charts the allocation of CRI's quarterly R&D budget by type of project. For example, new platforms account for a consistent $8 to $10 million per quarter, with a little more allocated in the next two quarters. The chart also includes the amount of the budget that is yet to be allocated, providing Chris with a rough indication of available capacity.

CRI Investment by Type of Project

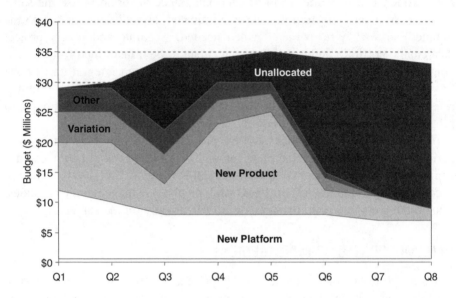

Chris explained how this analysis is put together, and how he uses it. "The chart comes from the actual approved budgets for all of our projects. I set a guideline for this chart to include approved budgets for approved phases of projects, as well as preliminary budgets for yet-to-be-approved phases.

"However, I also screened-out planned budgets for any project not yet through its Phase 1 approval. Until a project goes through its Phase 1 approval, the budget estimates are usually not accurate, and we eliminate a number of projects by the Phase 1 review. So, for example, the budget for a project we approved at Phase 1 earlier today is now included.

"I also have the ability to look at this same chart and include all budgets from all projects if I want. And I can even include projected budgets from planned products not yet in development. This can be a useful preview, but I tend not to rely on it.

"The chart also updates automatically as time progresses. As we move through Q1, for example, some of the data will be based on actual costs, and, at the end of the quarter, the data will all be actual costs."

Chris showed how he could go back in time and do this chart for the four previous quarters based on actual data, which is the extent of the historical data created so far. He then described how he uses this analysis. "By looking ahead I can see that we have some development capacity freeing up in Q3, but much of this is expected to be used in Q4 and Q5, because we have four large, new product efforts that are ramping up their Phase 2 work at that time."

He then re-did the analysis to include all planned projects and projects in Phases 0 and 1. "There is not much of an impact expected in the next four quarters, based on these projects, but we'll need to take some actions. I'll meet with some of the project managers and see if we can accelerate some of their projects' Phase 2 work from Q4 and Q5 into Q3 in order to level the demand more. Then I'll meet with our product managers to see if we can start work on some of the new products by Q3, although we may need to go slowly after that because of the constraints in Q4 and Q5. In any case, we'll need to work all of this down to the resource level, but I try to adjust the big picture, our strategy, and then let others adjust this as needed to get the best resource utilization."

Chris also talked about the difference between dynamic portfolio management and the periodic portfolio management CRI formerly practiced. "Previously, we did portfolio analysis like this periodically, and I had to have someone manually collect all of the data, which, of course, nobody wanted to do. It took three to four weeks to put together a chart like the one I'm showing you, and then I asked to look at it a little differently, like I just did. Of course, they hadn't collected what I wanted, so we had to spend another couple of weeks doing that. Then, when I went back to look at the first chart again, it was so out of date it wasn't useful."

With a big smile he said, "I like this a lot better."

Chris Taylor's example is just one of many cases where dynamic portfolio management is much more effective and efficient than periodic portfolio management. The advantages can be multiplied many times over for all the different types of portfolio analysis that are needed.

On-Demand Portfolio Analysis

This next topic of portfolio management is one that I call *on-demand portfolio analysis*. This means the ability to create a portfolio chart from a library of charts, with any selection of data compared to any other selection of data, whenever it's needed.

On-demand portfolio analysis requires integration of the portfolio analysis system with all project data. This means that everyone is able to do on-demand portfolio analysis, and all project teams must be part of a common system, such as a DCM system. On-demand portfolio analysis requires the transition of portfolio analysis from a technique and tool such as a spreadsheet to an integrated enterprise system.

Most people have found that, if portfolio analysis is to be helpful, it must provide multiple ways to understand a portfolio of projects. From an overview analysis, a portfolio manager may want to take a "deep dive" into something that he's concerned about. He then may want to look at the analysis slightly differently. Portfolio analysis can be done on dozens of types of data, different time periods for the data can be selected, a wide range of comparisons can be made, and multiple charting techniques can be used to analyze the information selected. In total, there may be more than 100,000 combinations of data used in portfolio analysis, and this cannot be done with a spreadsheet.

On-demand portfolio analysis also makes portfolio analysis available to all senior managers involved in guiding R&D. They no longer need to ask someone to collect data and analyze it in a certain way. They simply request the analysis they want from their DCM system. Most senior managers find that there are specific portfolio analysis charts that they want to look at regularly, and they build a personal library of charts that they can access quickly.

We saw Chris Taylor using on-demand portfolio analysis in the previous example. He looked at one chart that he regularly reviewed to monitor how the investment in R&D was being allocated and what was available to invest. Based on what he saw, he then looked at the picture a little differently with different data in order to determine the actions he needed to take.

Integrated Portfolio Management Process

When portfolio management is transformed from a tool to an integrated system, it can then become a more formal management process, and this is what companies are doing in the next generation of product development. This enables a company to establish clear responsibilities and steps for portfolio management, as well as consistent timing. The portfolio management process is integrated with the phase-based product-approval process and other strategic processes. In Chapter 12 we discussed how some companies may choose to combine the phase reviews for new products, enabling them to make go/no-go decisions for more than one product at a time in order to make resource allocation decisions across projects. This simultaneous phase review of multiple projects also enables companies to incorporate regular portfolio management decisions at the same time.

Most companies will individually tailor the portfolio management process because it needs to fit the nature of the products the company develops and accommodate other company-specific processes. Let's look at how CRI integrated its portfolio management process.

Like most companies of the previous development generation, CRI used a phase-based decision process to approve new product projects phase by phase. This worked very well and accelerated time to market. Nevertheless, CRI began to realize the limitations we discussed earlier, and began to do periodic portfolio analysis whenever some particular concern arose about the mix or timing of projects.

With its transformation to the new practices and systems of the R&D Productivity Generation, CRI adapted its phase-based decision process to incorporate portfolio management. In addition to project phase reviews, its product approval committee (PAC) began meeting quarterly to review its portfolio of projects, and also started performing dynamic pipeline management, which we'll discuss shortly. Eventually, the PAC started to make all decisions to start new projects at these quarterly meetings, since it had its best understanding of resources at the time of these meetings. This quarterly timing of proposed project reviews required a little adjustment on the part of those who wanted to start new projects, but eventually it worked out well, and even provided an inducement to those who wanted to launch new projects to submit their formal proposals in time for review at the next quarterly meeting. This deadline pressure is helping accelerate new projects at CRI.

The company subsequently included Phase 0 and Phase 1 reviews and approvals at these quarterly PAC meetings in order to gain more visibility of resource allocation, because once projects are through a Phase 1 approval at CRI they are rarely canceled. However, the PAC found that too much time was required for all Phase 1 reviews to be done effectively at the same time, so it went back to doing Phase 1 reviews individually, but looked at anticipated Phase 1 decisions during the quarterly portfolio meetings.

Chris Taylor explained how the new DCM system enabled portfolio and pipeline management to evolve into a process, not just a technique. "Once we could get portfolio data on demand, instead of weeks after asking for it, we began using it at every phase review. Then we found that we needed to spend time at phase reviews talking about projects other than the one we were reviewing, and this distracted us from the project we were supposed to be focusing on. So we started meeting quarterly to do the portfolio reviews. Finally, we published this agenda to the entire organization to set broader expectations and fit it into other activities in the business. After all this, we had a new process."

DYNAMIC PIPELINE MANAGEMENT

When companies first started attempting portfolio management, they found that pipeline management was also critical. As we defined at the outset, portfolio management aligns development investments with the business strategy, whereas pipeline management aligns resources with development investments.

Pipeline management requires even more data than portfolio management because it uses resource information to balance capacity against demand. As we discussed in Chapter 6, on the subject of medium-term capacity planning, resource management provides the capacity information needed for pipeline management. Right now, we're looking at how this resource information is combined with project information to balance the product development pipeline.

In Stage 1, periodic pipeline management was frequently simplified by determining project priorities and then allocating resources down this list of priorities until resource constraints were reached. While this is still a worthwhile technique, dynamic pipeline management improves pipeline decisions with pipeline modeling and optimization techniques.

The term *dynamic pipeline management* is used in the same way that it was used for portfolio management. All of the information for pipeline management is continuously available and always up to date. Executives responsible for managing the product development pipeline can get a snapshot of the pipeline at any time. Moreover, they can also simulate a range of alternatives to help them make decisions.

Real-Time Integration with Project Resource Management

In Stage 1 pipeline management, companies manually collected project information such as the phase a project was in, expected financial results of the project, and so on. They also profiled the resource needs of each project, a time-consuming task. A single project could have dozens of people working on it, with different skills, and those dozens of developers could end up putting in very different amounts of time over the length of the project. All these possible variations, in all their possible combinations, could require the collection of hundreds or thousands of pieces of resource data.

Companies at the first stage of pipeline management tried to combine this mass of resource data with portfolio information that contained the critical attributes of each project, and to enter all of this information into a spreadsheet to do pipeline management. As with periodic portfolio management, the data used in periodic pipeline management were frequently out of date by the time they were all collected. Understandably, this cumbersome process was only performed when there seemed to be a particularly burning need to "do" pipeline management.

Dynamic pipeline management uses data already available from other DCM systems. As illustrated in Figure 18–2, dynamic pipeline management is

Dynamic Pipeline Management

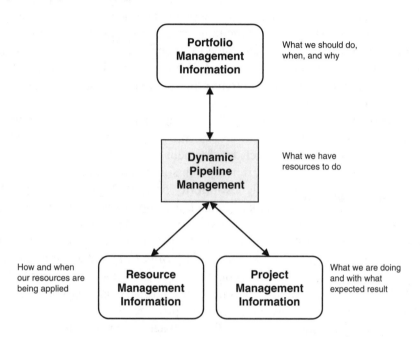

at the center between two layers of information. Essentially, it performs the demand/supply balancing between resource requirements and resource availability. Dynamic pipeline management is based on real-time integration with resource information, project information, and portfolio information, which requires these systems to be highly integrated.

In Chapter 6, we looked at the tasks of keeping track of all resource assignments and extending the visibility of development resources to include the expected future resource needs of all projects. This provides the demand information. We also discussed how the horizon of resource demand information can be extended by the introduction of resource estimates for planned projects. Resource supply information is also provided by the resource management system. This identifies what resources are available.

Dynamic pipeline management provides a set of integrated tools that enable all of these data on resource supply and demand to be modeled. A company can optimize its pursuit of a critical strategic objective, such as revenue or profit from new products, by identifying the appropriate product portfolio and the appropriate priority of the products in the development pipeline.

Pipeline Simulation and Optimization

As was previously mentioned, companies in the initial stage of pipeline management usually started out making their pipeline decisions by establishing project priorities using a set of ranking criteria or a predetermined scoring scheme. Resources were then allocated to projects according to their ranking, until resource constraints were reached. While this certainly helped to better direct resources to the most important projects, it still fell short of optimizing revenue and profit from new products.

The objective of dynamic pipeline management at Stage 2 is the optimization of revenue and/or profit from new products. This is achieved with the optimal output from the product development pipeline. In most cases, this is different from simply allocating resources to the highest-priority projects. With dynamic pipeline management, a company can simulate or model alternative mixes and timing of projects in the development pipeline, and estimate the projected revenues and profits from various mix and timing combinations. Let's look at an example of how one of the business units at CRI set about optimizing its pipeline output.

> With dynamic pipeline management, a company can simulate or model alternative mixes and timing of projects in the development pipeline, and estimate the projected revenues and profits from various combinations.

CRI simulated a range of alternatives for the portfolio of projects in its development pipeline. This simulation, illustrated in Figure 18–3, automatically determined the expected revenue profile for each portfolio alternative, and matched each alternative against the feasibility of meeting its resource requirements. These simulations allowed the company to construct a project portfolio with the optimal revenue profile for the resources available.

The resource profile chart depicted in Figure 18–3 shows the expected resource utilization by period for one portfolio alternative. The bars represent the resource demand for that portfolio by period (in this case, months), and total resource availability. In this example, resource requirements include current assignments and estimated future resource needs. In the portfolio alternative illustrated, resource needs exceed availability by a little in periods 2 through 4, but management thinks that this can be resolved with some scheduling adjustments, without affecting project completion dates.

The portfolio alternative just depicted is only one of more than 40 that CRI considered, and each one created a different resource utilization profile. Each alternative was plotted against the same resource profile, since resource

F I G U R E 18–3

Pipeline Simulation at CRI

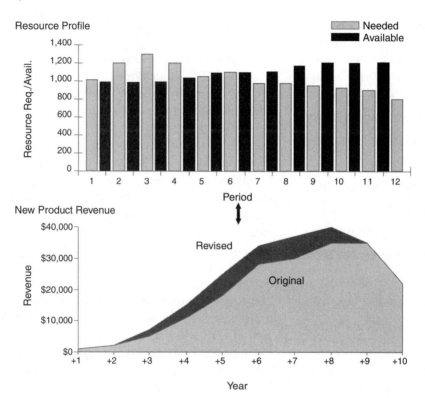

capacity is considered fixed for the purpose of this analysis. Of all these alternatives, only 12 fit the profile of available resources.

The new product revenue chart in Figure 18–3 shows the revenue profile for the depicted portfolio alternative (labeled "revised"), compared to the original baseline plan (labeled "original"). This pipeline alternative is expected to increase revenue in years three through nine, with an overall revenue increase of almost $30 million over the current baseline plan. To get to this alternative, CRI evaluated each of the 12 portfolio alternatives that fit available resources, and found this to be the most favorable.

When CRI performs its simulations of alternative portfolios, it looks at the impact of each portfolio on all capacity planning skills. Usually, only a few of these are constraining skills, and it focuses on these for the optimization.

Chris Taylor discussed the impact of this ability to test alternative portfolios through simulations. "We can now iterate dozens of portfolio and pipeline alternatives to optimize revenue. Sometimes, the optimal alternative is one we would not have intuitively considered. Overall, I think that we have increased revenue by tens of millions of dollars using this new capability. And this helps us achieve our strategic objective to get more output from our relatively constant investment in R&D."

Sometimes CRI also did pipeline simulations using probabilistic-weighted revenue and profit forecasts. These simulations projected lower revenue but generally did not change the optimum portfolio. Nevertheless they gave CRI more confidence in its decisions.

Pipeline management can now be integrated as a regular management process with portfolio management and the phase-based decision process. In fact, it becomes the link between the two.

CRI is expanding its pipeline optimization even further by looking at pipeline alternatives beyond simply including or excluding projects. It is requiring that all projects identify a project plan that is slower than its approved baseline plan and another that accelerates development. Guidelines are established for creating the alternative plans so that they take into account more or less of predefined resource constraints. Each alternative plan also changes the projected revenue and profit, based on the acceleration or delay in completion. CRI will thus be able to simulate not just the inclusion or exclusion of projects, but also alternatively paced project schedules, to determine the optimal revenue output. Instead of simulating approximately 40 portfolio alternatives, it could simulate hundreds of variations on those alternatives, and find an even better solution.

Integrated Pipeline Management Process

At this stage, dynamic pipeline management can also become a management process, not just a periodic technique. Just as with dynamic portfolio analysis, dynamic pipeline analysis can now be performed on demand. Those responsible for pipeline management can do the analysis with current data any time they want, and easily analyze dozens of alternatives.

Pipeline management can now be integrated as a regular management process with portfolio management and the phase-based decision process. In fact, it becomes the link between the two.

We previously discussed how CRI did pipeline management in its quarterly portfolio review to determine the feasibility of portfolio alternatives. At its four quarterly meetings, the PAC usually made adjustments to the pipeline to accommodate whatever problems had recently arisen. Once, it had to make some major changes as a result of a setback in a critical project. In addition, the strategic group at CRI generally did a major review of the pipeline once a year, in conjunction with defining product strategy as part of its annual financial planning process. During this annual pipeline review, the group reconsidered its entire product development pipeline and typically made strategic changes. We'll look at this more closely in Chapter 20.

REQUIREMENTS

Dynamic portfolio and pipeline management requires an integrated DCM system because it utilizes data maintained in other DCM systems, specifically resource demand and capacity data from resource management and project data from project management. Without this integration, it's difficult for a company to progress beyond sporadic portfolio and pipeline management.

In addition to the systems required, a company has to define the portfolio and pipeline analysis expected to be sure that the system is creating the necessary information and that senior managers are using it consistently. In conjunction with this, most companies establish a library of portfolio and pipeline charts.

Finally, a company needs to revise its management processes to take advantage of dynamic portfolio and pipeline management as a regular management capability, similar to what we saw at CRI. In many cases, this will involve changing a portion of its phase-based product-approval process.

BENEFITS

Dynamic portfolio and pipeline management activates some of the overall benefits of the next generation of product development.

1. New product revenue and profit can be increased through the optimization of the new product pipeline beyond what was possible by simply allocating resources to project priorities.

2. Resource utilization can be further increased by optimizing the development pipeline.

3. Dynamic portfolio and pipeline management provides more control and visibility over product development, which in many cases will lead to different and better decisions on project priorities and project portfolios.

4. Manual collection and analysis of data are eliminated and replaced with automatic, on-demand portfolio and pipeline analysis.

SUMMARY

The initial stage of portfolio and pipeline management began toward the end of the previous generation of product development, and it introduced some important, even if basic and periodic, improvements to the process of setting priorities across projects. Dynamic portfolio and pipeline management enables an entirely new stage of management, addressing the limitations of the previous stage and introducing some exciting new capabilities. On-demand portfolio analysis, made possible by the on-demand availability of current data, opens up a broader understanding of product portfolios. Pipeline simulation enables optimization of revenue and profit from new products beyond what was possible by simply setting project priorities.

19

CHAPTER

Comprehensive Financial Management of R&D

R&D is one of the largest, if not *the* largest, investments in many companies, and one of its most valuable assets. Yet to date, too many companies have inadequate financial management practices. This is something that will change in this new generation of product development. Chief financial officers (CFOs) are taking increased responsibility for the financial management of R&D.

Let's look at a couple of the more common failures in the financial management of R&D. The financial information used in long-term financial planning, which is usually part of a company's formal strategy, is often misaligned—if aligned at all—with the information used for reviewing and approving investments in new products. Financial information, such as projected revenue, capital investment, and profit, is a critical ingredient of the decision to develop any new product. At the same time, most companies include the projected revenue from products currently under development in their long-term financial plans. While one would naturally assume that the projected revenue from new products would be based on a consolidation of the revenue projections used to justify individual new product investments, that is not the case at most companies. For example, a company may be strategically planning on new product revenues of $20 million per year, but the expected revenue from products under development may be only $12 million. And the company may not even know that this $8-million discrepancy exists.

Another serious problem of financial misalignment concerns the two types of R&D budgets: project budgets and functional budgets. As part of the annual planning process, the R&D budget is established through the individual *functional budgets* of the resource groups involved in R&D. *Project budgets* are used to distribute this R&D investment among individual projects, typically through a phase-based approval process. Yet most companies fail to reconcile these two budgets. A company could budget $100 million in R&D across many functional department budgets, and not know if it has allocated $40 million or $140 million, simply because there is no system or process to reconcile the two types of budgets.

Imagine how a board of directors would react to an audit report that identified these two financial breakdowns:

1. Long-term financial plans for new product revenue and profit are unreliable because they aren't based on actual investments in new products, and this creates a material risk to future financial expectations.

2. R&D budgets are out of control. Functional budgets may be significantly over- or underallocated because they are not reconciled with the project budgets to which the functional budgets are supposedly allocated.

While most companies live with these financial control risks because they have always lived with them, boards of directors, post Sarbanes-Oxley Act, have higher expectations of financial control.

In this chapter, we will look at the expanding role of the chief financial officer in the next generation of product development, and how better financial management of R&D will be enabled by the DCM systems and management processes of this new R&D Productivity Generation. In Chapter 13, we looked at integrated financial planning and project budgeting for new products. This was the project-level view of R&D financial management. Here, we will focus on the consolidated view. We'll also examine the new role of the CFO in R&D financial management. Financial executives may find it useful to look at these two chapters together.

THE CFO'S NEW ROLE IN R&D

Over the past few decades, financial executives have increasingly broadened their role in the overall management of their companies. Initially, their focus was on basic control of accounting, including accounts payable, accounts receivable, general accounting, and so forth. Systems and processes introduced in the 1970s and 1980s enabled new financial management practices for basic accounting. Financial executives introduced new financial management practices in manufacturing during the 1980s and 1990s, with the advent of enterprise resource planning (ERP) systems. New financial management practices were also introduced in backlog management and revenue recognition during that period.

Over the last few years, CFOs have come under increased pressure. There is much less tolerance of ignorance of the R&D budgeting process. Boards of directors, and even shareholders, are taking precipitous action with the benefit of hindsight when there are surprising negative results in a business. Financial control of R&D will undoubtedly come under increased scrutiny by auditors and boards.

One of the common themes in the improvement of financial management practices is the need for enterprise-wide information systems to produce the financial information necessary for financial control. I believe that the lack of these systems is the primary reason why most companies have inadequate financial control of R&D. Let's look at how one CFO is increasing the financial control of R&D in our hypothetical company, Commercial Robotics Inc.

> The lack of enterprise-wide information systems is the primary reason why most companies have inadequate financial control of R&D.

Shaun Smith, the new CFO at CRI, is a farsighted financial leader brought in by the board of directors to control the company financially during its expected rapid growth. Since R&D is the most critical process at the company, Smith ordered the following specific financial management changes in R&D:

- All business-level R&D financial information, including the new product projections used in strategic planning, will be based on consolidations of the financial information used by individual projects.
- New product project budgets will be reconciled with R&D functional budgets to determine how much of the R&D investment has been allocated. We will use a "charge-out" approach to functional budgets.
- Monthly financial reporting of R&D will be used by the company's executive committee to manage R&D.
- All projects will use standard financial models and approved financial assumptions in order to ensure consistency in financial evaluations of projects.
- Finance will determine the cost of R&D resources so as to fully charge out all R&D costs.

We discussed the last two of those items in Chapter 13, so we'll focus here on the first three. With the implementation of a new development chain management system, Shaun requires that all new product development project teams use standard financial planning models. This will enable the consolidation of financial information across all projects. With the introduction of "standard cost charges" for all R&D resources, CRI is now able to reconcile functional budgets to project budgets.

CONSOLIDATION OF NEW PRODUCT FINANCIAL INFORMATION

The pieces of new product financial plans are scattered throughout most companies, and there may even be multiple versions of the same plans. This makes it very difficult for anyone to collect and consolidate the information that is absolutely critical to decision making and control. This state of disarray is comparable to someone running an investment fund and not keeping accurate records of how much is invested, where it's invested, and what the return is from those investments.

It is widely—and wrongly—assumed that because certain financial information is available in the company, it is being used. In fact, the information can't be consolidated because it is so widely dispersed: It resides in desktop computer files, in spreadsheets, in presentations, and who knows where else. You may recall that in Chapter 13 we reviewed integrated financial planning and project budgeting for new products, and explained that one of its benefits is to consolidate financial information across all projects. Here we will focus on the use of this consolidated financial information in financially managing R&D. Financial information is maintained for each project using the standard DCM financial planning models discussed in Chapter 13, and, since this information resides in an integrated financial system, it can automatically be consolidated at the business level. Examples of financial information consolidated at the business level include revenue projections, profit projections, capital requirements, planned operating costs, and investment return.

Trying to collect and consolidate this information manually is very difficult, except for smaller companies with only a few projects. Project financial information is frequently maintained in various formats, requiring conversion

> It is widely—and wrongly—assumed that because certain financial information is available in the company, it is being used. In fact, the information can't be consolidated because it is so widely dispersed.

to a standard business-level summary. It takes a lot of time to collect the information, and by the time it is all collected, much of it has changed. As a result, the business-level financial information consolidated from the project information is inaccurate.

In the next generation of product development, an integrated financial system creates business-level financial reports directly from the project-level source data. The financial reports are accurate at the moment they are created. The integrity of business-level financial reports is high because the reports are based on exactly the same financial information used by project teams and executive management.

Let's look at a few examples of how business-level financial reports are created and used.

Consolidated Revenue Projection

Commercial Robotics implemented integrated financial reporting as one of its priorities. Figure 19–1 illustrates consolidated revenue projections for CRI's Retail Robot business for selected years four through nine of its revenue plan. Note the inclusion of projected revenue from four products already released to the market (Market), a product in development (Development), the Fast-Food Robot I, and six planned products (Plan) whose development has not yet started.

Revenue projections for products already on the market can be derived from other sales forecasting systems, and then integrated into the financial system within DCM, or can be maintained directly for this purpose. Again, the revenue projections from products in development come directly from the project information in the DCM system. Revenue projections from planned products either can come from an integrated product strategy system (see next chapter) or can be maintained in the DCM financial system for this purpose.

The consolidated business-level revenue projection is based on the actual data currently being used in each of the individual projects. So, for example, if the revenue projection changed in the Fast-Food Robot project due to a different price estimate or quantity estimate by the project team, the consolidated business-revenue projection would be automatically adjusted. This automatic adjustment synchronizes financial information from the project with financial information at the business level and ensures that everyone is looking at the same set of facts.

Some companies will establish guidelines for "freezing" or taking snapshots of consolidated revenue forecasts at specific points in time. For example, CRI might take a snapshot of the consolidated revenue forecast when a major project, such as the Fast-Food Robot, completes a Phase 1 review. A company

F I G U R E 19–1

Consolidated Revenue Projection for the Retail Robot Product Line at CRI

Consolidated Revenue Projection

Retail Robots	Status	Years					
		4	5	6	7	8	9
Early Robots							
Dispenser Robot	Market	$25,000	$20,000	$5,000			
Take-Out Robot	Market	$45,000	$35,000	$15,000			
Ticket-Taker Robot	Market	$47,500	$45,000	$20,000			
Ticket-Taker Robot	Market	$47,500	$45,000	$20,000			
	Early Total	$165,000	$145,000	$60,000	$0	$0	$0
Fast-Food Series							
Fast-Food Robot I	Development	$8,000	$44,500	$99,360	$158,025	$0	
Fast-Food Robot II	Plan			$66,240	$168,725	$365,400	$200,000
Fast-Food Robot III	Plan					$141,350	$350,000
European FFR II	Plan						$5,000
Asian FFR II	Plan						$2,500
	Fast-Food Total	$8,000	$44,500	$165,600	$326,750	$506,750	$557,500
New Retail Series							
Retail Robot I	Plan		$12,000	$40,000	$50,000	$20,000	$10,000
Retail Robot II	Plan				$30,000	$90,000	$140,000
	New Retail Total	$0	$12,000	$40,000	$80,000	$110,000	$150,000
Retail Product Line Total		$173,000	$201,500	$265,600	$406,750	$616,750	$707,500

may take quarterly snapshots and ask every project team and marketing group to check and update its projections at the end of every quarter. Trends in revenue forecasts can then be compared from quarter to quarter. Certainly, most companies will want to establish a baseline revenue projection consolidation annually as part of their business planning.

The consolidated revenue projections can be used in long-term planning in several ways. Some companies may want to discount the consolidated revenue projection prior to establishing a formal projection in the long-term financial plan. At CRI, the manager of the retail robot business reduced the projected revenue increase by 10 percent cumulatively for planned products in order to be more conservative. He could do this because the consolidation showed a significant increase in revenue.

The manager of another CRI business had the opposite problem. His consolidated revenue forecast showed a flat revenue projection, which was a problem because every business was expected to grow by at least 15 percent per year. His business's investment in R&D was not expected to yield any growth, and this wasn't acceptable. Either the wrong projects were in development, or the product strategy was weak, or the market was declining. Since independent market forecasters were predicting 10 percent annual growth in his market, the business unit manager figured it had to be one of the first two, and called for a complete evaluation of the product strategy, including products in development. This resulted in a significant change to the product strategy for that business. The manager later commented on this episode. "Without the new visibility of consolidated revenue projections based on project data, our marketing group would have projected revenue from the top down, mainly relying on market-growth estimates, and these would have been very different from what we were actually doing. We were able to identify this in advance and change our strategy to accommodate targeted growth."

Shaun Smith, the CFO, also consolidates revenue projections across all five of CRI's divisions, since in the end it's the growth of the overall business that is most important. This consolidation is illustrated in Figure 19–2. The consolidated revenue projections from each division are shown. Each of these division projections is based on the consolidation of revenue forecasts for products

F I G U R E 19–2

CRI's Consolidated Revenue Plan

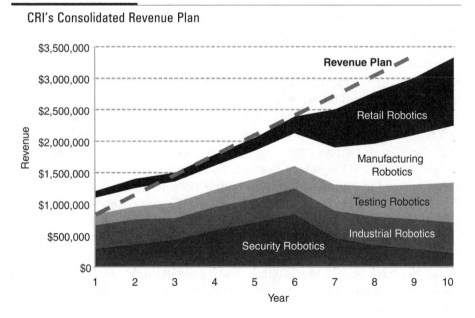

already in the market, projected revenue from new products currently being developed, and planned revenue from products not yet in development. As you can see, the Retail Robotics Division is expected to be the primary driver of CRI's growth over the next 10 years. Security Robots are providing some near-term growth, but this market is expected to mature in six or seven years.

The line on the chart represents CRI's 15 percent targeted growth. There is good coverage from the current product strategy, although Shaun is a little concerned about years three to five, since CRI expects to have a reasonably firm product strategy for the next five years. This is something he and the executive team at CRI will continue to watch closely.

Consolidated Capital Plan

CRI also consolidates all of its capital investment plans in order to get a view of total expected capital expenditures. This consolidation includes the capital requirements from all new products under development, longer-term capital forecasts from planned products, and all other capital spending plans. This requires integration not just of product development projects, but also of all current and planned capital investments.

To accomplish this consolidation, CRI extended its DCM system to manage all major projects outside of R&D, as well as those within R&D. It wanted to use a common system and process for all project-based management, so it implemented a similar system for capital projects, IT projects, and process-improvement projects. This system enabled consolidation of financial information across all projects as well.

Shaun Smith reviews the consolidated capital plan quarterly. Recently, it indicated projected capital spending that is well above CRI's total capital budget, which prompted an examination of the accuracy and timing of these projections. Shaun found that many of the capital investments could be delayed, and the consolidated capital plan was changed to reflect these delays. However, the plan still shows excess capital requirements above the overall capital budget, because CRI's growth is driving more capital spending than was anticipated before the company started doing this consolidation.

Shaun reacted to these findings by putting together a new capital plan for CRI's board of directors, including a recommendation for selling additional stock to secure the necessary capital. In his presentation to the board, Shaun said, "Without this new consolidation of all our capital requirements, we would not have been able to anticipate these capital needs sufficiently in advance. Now we have time to evaluate alternatives and raise capital when it's to our advantage."

Consolidated New Product Profitability and Expense

Some companies may also find it helpful to consolidate the projected income from new products as well as selected expense categories, in order to provide some visibility of expected changes to functions that are "downstream" from product development. For example, consolidated income from new products provides an additional insight into the future business beyond the consolidated revenue projections we just discussed.

CRI uses these consolidated projections selectively. For example, consolidated marketing and sales expense for new products is used by the VP of Sales to give him some insight into potential shifts in costs across sales channels. The VP of Services bases increases in the size of the customer support organization on the projected service revenue and expenses consolidated from new product plans. Without this, his planning would not be aligned with the product strategy.

The CFO also looks at the consolidated return on investment across all projects to see if the overall weighted average return on R&D investments is adequate. This weighted ROI went down substantially when CRI started fully costing all R&D resources, but it has since returned to appropriate levels.

Consolidated Project Budgets

Consolidation of individual project budgets can also provide some visibility of expected R&D investment. It can show how much of the total R&D budget is allocated and how much is still available. We looked at an example of this in Figure 18–1.

RECONCILIATION OF FUNCTIONAL AND PROJECT BUDGETS

As we have discussed, there are two types of R&D budgets. The first type is the annual budget prepared for all functions or departments (resource groups) in R&D, as well as for all other functions or departments throughout the company. These budgets tell functional managers how much they can spend on compensation (resource costs) and all other expenses, such as travel and training. Some functional budgets also include allocated costs for expenses such as facilities, and this can be helpful in managing the total costs of a function, as we will see in the example that follows.

The second type of budget is the project budget, which is approved phase by phase by the Product Approval Committee (PAC) or similar executive group. The project budget includes approval of resources and other expenditures for that project. Sometimes the actual costs of projects are managed against the approved project budgets, as was illustrated in Chapter 13.

In theory, the functional budgets are "allocated" to product development projects when the PAC approves a project, but, until this new generation of product development management, this could not be done in practice. There were no systems or processes to enable a reconciliation of what was allocated and what wasn't; hence my earlier example of the company that didn't know if it had allocated $40 million or $140 million of its $100 million R&D budget. The actual allocation takes place function by function, as we will see in the upcoming example.

Tracking Functional Budget Allocation to Projects

CRI uses what is called a "charge-out" process to reconcile its functional budgets with its allocation of resources to product development projects, so that it can fully allocate the R&D costs to approved projects. To do this, CRI first had to establish a full cost for all its resource assignments, as we originally discussed in Chapter 5.

> In theory, the functional budgets are "allocated" to product development projects when the PAC approves a project, but until the new generation this could not be done in practice.

Figure 19–3 illustrates this reconciliation for the Software Engineering group. The financial report shows the charge-out of resources assigned to projects compared to the costs for the group. In the current month, $243,510 was charged to projects using Software Engineering resources. This is computed from the number of days assigned by an individual to a project, multiplied by the daily charge-out cost for that individual. For example, all of the developers from Software Engineering working on the Fast-Food Robot project were charged out at $125,050 for this month.

The total cost for the group was $243,919.66 for the month. This was made up primarily of compensation costs, but also included such costs as equipment depreciation and charged overhead for facilities. Netting the total charged to projects against its costs for the month, Software Engineering ended up with a small, uncharged balance of $409.66 for the month. CRI sets the goal for all groups to charge out approximately all of their costs, and Software Engineering was pretty much on target. Year to date, the group had charged out $7,382.05 more than its costs, again pretty much on target.

CRI also reported projected project charges for the year, which included future assignments (those already committed and approved) and fees already

Charge-Out of Software Engineering Costs at CRI

Software Engineering Costs

Charged to Projects	This Month Actual	YTD Actual	Year Projected
Major Products			
Fast-Food Robot I	$125,050.00	$875,300.00	$1,450,000.00
Fast-Food Robot II	$12,200.00	$73,200.00	$77,300.00
Retail Robot I	$56,050.00	$336,300.00	$450,735.00
Hotel Cleaner	$12,500.00	$75,000.00	$93,200.00
ID Robot	$5,600.00	$33,600.00	$43,200.00
Other Projects			
Industrial Software V10	$5,210.00	$10,500.00	$21,000.00
Server Upgrade	$3,500.00	$7,500.00	$15,000.00
Material Tester V3	$6,700.00	$14,300.00	$28,600.00
Industrial Robot V2	$16,700.00	$45,200.00	$90,400.00
Total Charged	$243,510.00	$1,470,900.00	$2,269,435.00
Costs (Actual/Budget)			
Compensation			
Salaries	$105,000.75	$630,004.50	$1,260,009.00
Bonuses	$10,500.08	$63,000.45	$126,000.90
Medical	$16,800.00	$100,800.00	$201,600.00
Taxes	$7,350.05	$44,100.32	$88,200.63
Other	$4,200.03	$25,200.18	$50,400.36
Expenses			
Travel	$23,200.50	$139,203.00	$278,406.00
Meetings	$8,500.20	$51,001.20	$98,300.00
Misc. Software	$5,000.00	$30,000.00	$60,000.00
Training	$17,500.80	$105,004.80	$210,009.60
Recruiting	$5,000.00	$30,000.00	$45,000.00
Equipment Depreciation	$23,500.00	$141,000.00	$282,000.00
Charged Overhead	$15,000.00	$90,000.00	$180,000.00
Miscellaneous	$2,367.25	$14,203.50	$28,407.00
Total Costs	$243,919.66	$1,463,517.95	$2,908,333.49
Total Charged (Uncharged)	($409.66)	$7,382.05	($638,898.49)

charged to projects. These projected project charges totaled $2,269,435 for the year. This is compared to projected actual and budgeted costs for the year of $2,908,333, which shows approximately $638,898 to be covered with future assignments. It's not unusual to have projected costs that exceed assignments, since more assignments will be made throughout the remainder of the year.

Basing Functional Budgets on Expected Resource Requirements

The other financial management practice in addition to reconciling functional and project budgets is to align functional budgets with expected resource requirements when the annual budgets are prepared. Doing so will make it much more likely that the actual allocation will be closer to the functional budget. More importantly, it will guide the setting of functional budgets for the upcoming year, as we will see in the following example.

Annual planning of the functional budgets was always a frustrating experience at CRI. Every functional department claimed that it needed a big increase in resources in the upcoming year or it couldn't meet expectations. Shaun Smith provided some insight on what went on prior to his arrival at CRI. "As I've been told, every R&D department manager would budget big increases, usually 30 percent or more, and then these would need to be cut back to maintain the overall level of R&D spending required for a profitable financial plan. There was no rationale behind the requested increases, just a sense that more resources were needed to do all the work. Eventually, every functional manager realized the way the game was played. Your request was going to be cut back proportionately, so you needed to request more than you thought you needed."

This was one of the things Shaun changed when he introduced new R&D budgeting practices. He made sure that medium-term resource capacity management, which we discussed in Chapter 6, was integrated with annual financial planning and budgeting. Headcount plans in the functional budgets are now based on forecasted needs from the resource management system. Functional budget requests are supported by projected resource needs based on the consolidation of needs from that resource group across all projects.

As a result, while overall R&D spending increased by 15 percent, the increase was better aligned to the mix of resource needs. Some budgets increased by more than 20 percent, while others had no increase at all. Shaun realized that some functions were still overstaffed, but he decided to leave that issue until later, since there was already enough change and CRI was moving in the right direction, balancing its functional budgets on fact-based assumptions.

REQUIREMENTS

Since comprehensive financial management takes place at the business level of aggregation of project information, it requires an underlying financial infrastructure. Common financial planning and project budgeting must first be implemented across all projects. This is typically done using standard financial-planning models, as was described in Chapter 13. This project-level financial information then needs to be consolidated across all projects.

While in theory the integration of financial information can be done manually using spreadsheets to collect the data, in reality this is only possible at small R&D organizations with few projects. DCM systems with an integrated financial management capability can provide this consolidation directly from all project-level financial information. Reporting of this information is usually flexible and customized for each company. The DCM system also needs to be integrated with the financial systems in order to reconcile the functional budgets maintained in an accounting or ERP system with the project budgets maintained in the DCM system.

> While in theory the integration of financial information can be done manually using spreadsheets to collect the data, in reality this is only possible at small R&D organizations with few projects.

Finally, this integration requires a new financial management process for R&D. The consolidated financial reporting and budget reconciliation described in this chapter is new to most companies and will require a process for the collection, review, and reporting of this information. It should be anticipated that the initial financial information may not be accurate when first collected and reported, because the underlying financial information was not used previously for this purpose. But this is all part of the process of getting financial control of R&D.

BENEFITS

Comprehensive financial management of R&D enables the CFO and executive leadership to gain financial control over R&D. The benefits are significant:

1. Aligning the expected financial results from new products with the long-term financial plan ensures that what is planned is consistent with actual new product investments. This alignment can avoid a major strategic disconnect, where a company may think it will grow,

but will be surprised to find not only that its investments didn't get it there, but that it could have known this in advance.

2. Fundamental control requires that project budgets are reconciled with the functional budgets they are allocated from.

3. Basing annual functional budgets on fact-based resource-needs estimates enables the mix of R&D spending to be adjusted to the expected needs of the development pipeline, resulting in more new product output.

SUMMARY

We discussed project-level financial planning and project budgeting in Chapter 13. In this chapter, we looked at the higher level of financial management, which I refer to as comprehensive financial management of R&D. Some companies are already beginning to find that their lack of financial control of R&D is a major financial defect, and this weakness will eventually be exposed to auditors and boards of directors. CFOs have a significant interest in this new generation of product development because it will enable them to gain the financial control of R&D that they need.

We have reviewed two major financial management practices of the new R&D Productivity Generation: The first was the alignment of financial planning with the consolidation of new product investments. The second was the reconciliation of project budgets with the functional budgets upon which they are based. Together, they will connect product development with financial management as never before.

20
CHAPTER

Integrated Product Strategy Process

In 2001, when writing an updated edition of my book on product strategy,[1] I wrote about the critical role that product strategy plays in most companies. In particular, for companies that depend on a continuous stream of new products for their livelihood, product strategy is the most important element of their entire business strategy. In addition, and most importantly in the context of the new R&D Productivity Generation of product development, product strategy sets the direction for product development. While writing that book, I became convinced that product strategy must be tightly integrated with product development. Next to not having a formal process for product strategy, the biggest mistake companies make is not integrating their product strategy process with their product development process. Here again, until this new generation, there weren't any integrated systems to provide the "glue" to bind these processes.

Product strategy defines the plan for new products that a company intends to develop. It is obviously integrated with the company's overall strategic plan, but it also must be integrated in several other ways. A product strategy plan includes products to be developed in the future as well as products already in the development pipeline and those already in the market, and an integrated product strategy must take into account the current status and expectations of all of these. Product strategy also needs to be expressed in financial terms to be incorporated into the long-term financial plan, and this financial plan includes both products already in the pipeline and planned products not yet in development. Finally, to be feasible, a product strategy must take capacity limitations into account, and this requires a company to simulate alternative product plans against available development capacity.

A product strategy plan includes products to be developed in the future as well as products already in the development pipeline and those already in the market.

In this chapter we'll look at how the product strategy process integrates with product development as a key part of the next generation of product development. In particular, we'll see how product strategy is integrated with resource capacity management, which we discussed in Section Two, and with the dynamic pipeline management and comprehensive financial management previously discussed in this section. We'll start by looking at an overview of the product strategy process and system. Then we'll focus on the product line plan, which defines the products that are scheduled for development over the planning horizon. As we'll see, a product line plan is made up of *planned products*, and these are the essential elements used to integrate product strategy throughout portfolio management and resource management.

A SYSTEM AND A PROCESS FOR PRODUCT STRATEGY

In my book on product strategy, I explained that companies with a better product strategy process consistently create better product strategies. A sustainable product strategy goes beyond dependence on the instincts of any single leader or team; it requires that product strategy become an institutional proficiency. Some companies have been able to achieve this institutional proficiency, while others are still struggling to achieve it.

In striving to create a product strategy process, companies have encountered some serious procedural obstacles. The first is that, until quite recently, there were no information systems to support product strategy. The product strategy process was thus a manual one, and its data-intensive nature posed a real problem in that regard. Like many other manual processes today, the product strategy process was often not given the attention it needed. The second difficulty is related to the first. The product strategy process exchanged information with other processes, but there were no systems for performing this exchange. In the next generation of product development, there *is* a system for product strategy, and this system is an integral part of the DCM system we have been discussing.

Figure 20–1 illustrates the major processes within a product strategy process. The three darker processes in the figure are the product strategy process, while the others are processes that use product strategy information. Platform and product line planning are at the center of the product strategy process and system, where planned products are defined. An effective and realistic product line plan cannot be created in a vacuum. It must be tested against available resource capacity through integration with pipeline management, and it must achieve expected financial targets. Product line planning must be integrated with these two systems and processes.

Platform and product line strategy are also highly integrated with the overall business strategy process, since it is a key element of strategy, but an examination of platform and product line strategy is beyond the focus of this book.

In addition to product line planning, there are two other important processes within the product strategy process. Idea management is the process through which a company collects and examines new product ideas from a wide range of sources. The process for developing a technology roadmap is also an important one, particularly for technology-driven companies.

F I G U R E   20–1

Major Processes in the Product Strategy Process

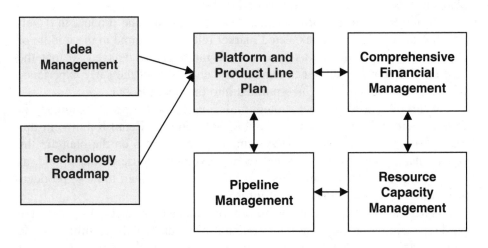

PRODUCT LINE PLAN

A product line plan is a time-phased conditional plan for the sequence of developing the new products of a product line that is itself the result of the product strategy process. There are several important elements of this definition. The product strategy determines the sequence in which products are developed and released. This sequence is time-phased throughout the lifecycle of a product line. Finally, the product line plan is conditional in that it can change in response to changes in market conditions, competitive factors, or resource availability. A product line plan may also be called a product roadmap.

Let's look at the example of the product line plan for CRI's Retail Robotics division, as illustrated in Figure 20–2. The division has three product lines, and the plan covers a 10-year horizon. The top two product lines are derived from the same product platform. The product line at the bottom includes the four currently marketed robots in what CRI refers to as its Early Robot Series. There are no new planned products for this product line. The rest of the product line plan is comprised of planned products. In the figure, CRI uses a notation scheme in which the lighter portion of each planned product indicates the development cycle (with the approximate phases marked) and the darker portion indicates the expected lifecycle of the product once it is released to market. For example, the Fast-Food Robot I product is currently in development in Year 1 and is planned to come to market in Year 4. From its release, it is expected to have a market life of a little less than four years.

The Fast-Food Robot II is planned to go into development at the beginning of Year 4, and to be released to market in the middle of Year 6. Once released, this product is expected to cannibalize the Fast-Food Robot I, which is planned to be phased out by the second half of Year 7.

In the Retail Robot product line there is currently one product in development in Year 1, with an expected market release sometime in the middle of Year 5. Development of this product was intentionally delayed so that the Fast-Food Robot I could get to market first and some of the early experience from that product could be incorporated into the Retail Robot.

With integrated product strategy, information on the products already in development in Year 1 (the Fast-Food Robot I and the Retail Robot I, in this example) comes from the DCM system. The phase dates on the plan are the actual phase dates from the project teams developing those products, and, as we will see, the resource and financial information attached to these products also comes from the actual project information.

There are also interrelationships among planned products. As part of the product planning process, these interrelationships can be defined similar to the way interrelationships were defined among the steps within a project. One of

CRI's Retail Robot Product Line Plan

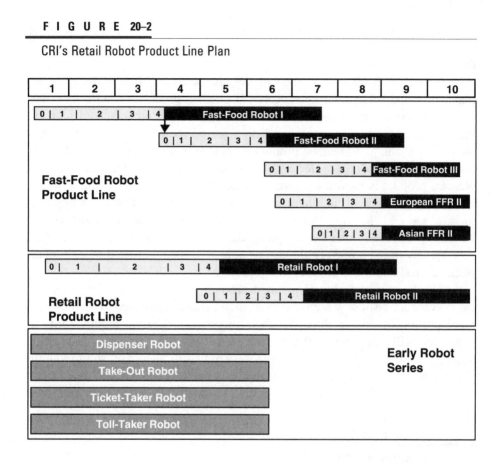

these interdependencies is depicted in the chart. It indicates the relationship of the Fast-Food Robot I to the completion of Phase 0 of the Fast-Food Robot II.

Planned Products

A planned product is a new product that is not yet started in development, but is formally on the planning horizon to be developed. Generally, each planned product is represented on the product line plan by an approximate date when development is anticipated to begin, an estimated duration of development, and an expected lifecycle once released to the market.

Planned products are the essential element of product strategy, similar to the way that project steps are the essential element to a project plan and projects are the essential element in portfolio management. When product strategy is integrated into a DCM system, planned products are "active," meaning that they have associated characteristics such as the expected revenue forecast. And, once a planned product is active, the information on it can be used throughout the DCM system.

Figure 20–3 illustrates some of the characteristics that make an active planned product. The Fast-Food Robot II planned product has a development cycle defined by the duration of the development period indicated with the lighter shading, and a revenue-generating period indicated by the duration of the darker shading. A planned product is created based on an estimated development period and lifecycle, and, if either of these is changed, then the planned product is moved forward or backward in the time schedule. The development schedule at the phase level indicates when each phase is expected to start and end. Moving this planned product forward or backward in time would automatically change the phase-date schedules. In addition, the cycle time for any of the phases could also be adjusted, and this would reposition the planned product on the product line plan.

The Fast-Food Robot II planned product also has several important attached characteristics. A high-level resource profile is attached to the development portion of this product, which estimates the resource requirements by phase for skill categories that are most useful for medium- and long-term resource planning. These resource estimates are tied to the phase, so moving the phase dates would, as we just saw, reposition the resource needs and project budgets into different months for planning purposes. We'll see how this infor-

F I G U R E 20–3

Characteristics of a Planned Product

Planned Product

| High-Level Budget | High-Level Resource Profile | High-Level Financial Estimates | Planned Revenue |

mation is used shortly. In addition, a project budget, again by phase at a high level of aggregation, can also be attached.

Financial estimates for the new product, including a revenue profile, are attached to the revenue portion of the Fast-Food Robot II planned product. These estimates are based on the starting date for the release of the Fast-Food Robot II product, and any change in that date will shift the timing of revenue and financial projections accordingly. Minor shifts in time of release could simply move forecasted revenue forward or backward, but any major change would most likely require new forecasts.

Planned product characteristics (the attached information) are used throughout other systems and processes in an integrated DCM system. In Chapter 6 we talked about medium-term capacity planning and how the introduction of planned products into medium-term capacity planning enabled an important extension in the resource-planning time horizon. It's these characteristics of planned products that provide the resource information necessary for this extended horizon. Likewise, longer-term capacity planning is dependent on the time horizon for planned products. In Chapter 18, where we looked at dynamic portfolio and pipeline management, we made reference to the inclusion of planned products. And in our discussion and examples in Chapter 19 on comprehensive financial management, we used the financial information attached as characteristics of planned products.

Standard Planned Product Profiles

When a planned product is created during product line planning, the characteristics we just discussed are also defined. While this may be somewhat time-consuming, the broad use of planned products throughout product development makes it worth the time.

In addition, some companies may choose to define standard profiles for planned products in order to simplify the work needed to create planned products. These profiles include standard characteristics such as phase durations, resource estimates by phase, development budget estimates, etc., for the development efforts of different types of products. For example, there could be standards for a simple product modification, a major product in an existing product line, a new product platform, a major technical upgrade to a product, and so on. When putting together a product line plan, these standard, planned product profiles can be used as the starting point, automatically adjusting the characteristics for the planned start time, which in turn can be adjusted as necessary.

Similarly, standard, planned product profiles can also include typical revenue and financial estimates. The standard then provides the starting point for the product being planned, and these financial estimates can also be adjusted. Standard, planned product profiles can be created for various types of products, such as minor products, major products, and revenue-enhancing upgrades of existing products.

In addition to making the product planning process more efficient, standard, planned product profiles can also increase the consistency of product strategy. At this point, so early in the thinking about a new product, estimates of development costs and sales revenues are tentative at best, but it's still valuable to have some estimates to work with at the beginning. These cursory estimates can be refined as the product moves into development. The point here is that starting out with a consistent set of standards helps focus thinking on the anticipated resource needs and financial estimates of each planned product.

Simulating the Feasibility of a Product Line Plan

When product strategy is integrated within a broader product development process and DCM system, as it is in the next generation of product development, alternative product line plans can be tested against capacity to determine their feasibility. This feasibility testing makes the resulting product line plan much more realistic.

Since planned products and current projects compete for resources, any capacity feasibility analysis must include both. By adding planned products to the development pipeline, alternative product line plans can be simulated against capacity, and the feasibility of each alternative plan can be tested. At the same time, the expected financial outcome, such as revenue by year, can be projected for each alternative. Once the set of feasible product line plans is determined, the financial outcomes can be compared and the most attractive plan selected.

We already saw an example of this pipeline simulation and optimization in Figure 18–3. With a fully integrated product strategy process, the ability to examine alternative product line plans is added to this simulation and optimization.

IDEA MANAGEMENT

If product strategy is considered to be the front end of product development, then idea management is the front end of product strategy. The product strategy process of deciding which new products to develop and when to develop them begins with ideas for new products or improvements to existing products. Traditionally, most companies manage new product ideas informally, if at all. In fact, many companies may find the very term *idea management* to be quite foreign. But new product ideas are very important, and so is their management, so let's look at how these ideas can be better managed in the new R&D Productivity Generation.

In her excellent book *New Ideas About New Ideas*,[2] Shira White vividly illustrates the importance of ideas. New ideas are ultimately behind every new product, new business, and new market. Here we will look at how ideas can be better collected, developed, and acted on more quickly in this new generation of product development.

> If product strategy is considered to be the front end of product development, then idea management is the front end of product strategy.

Figure 20–4 illustrates the extended product development pipeline with product strategy in the form of a product line plan in front of the phases of the product development process, and idea management preceding that. We just reviewed how planned products are incorporated into the product development pipeline to extend visibility of resource needs and consolidated financial projections. *Idea management* is the process for gathering ideas from a variety of internal and external sources, organizing them, and prioritizing them to further the business strategy. It extends product strategy even further.

Idea management is both a system and a process. It encourages the submission of ideas by making it easy for anyone to submit an idea, and it collects all of these ideas in a common system. Once in an idea management system, all ideas can be immediately categorized and made available for analysis and comparison. Ideas can also be enriched with supporting documents and links in a web-based system.

An idea management system and process also ensure that all ideas are consistently evaluated. They can then be incorporated into product plans, inventoried for future opportunities, or disposed of. As an idea progresses through an idea management system and process, more detail is added to it. This makes idea management a way for broad collaboration around ideas, which can have tremendous benefits. To quote Shira White, "Nothing is created in a vacuum. Innovation comes from connectivity, intersection, and the integration of ideas. We all draw on our own experiences, feelings, emotions, and perceptions. As we connect with others, these come together to form new patterns. Ideas merge to form new ideas."[3]

When idea management is part of a fully integrated DCM system, ideas can evolve into planned products or specifications for product enhancements.

F I G U R E   20–4

Extended Front End of Development Pipeline

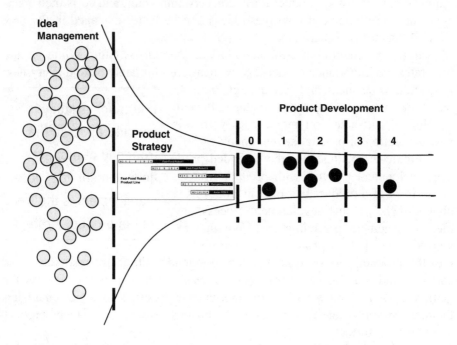

As an idea progresses through the system, it picks up more characteristics, perhaps even reaching the point where those characteristics include development estimates, revenue projections, and a planned project-launch date.

CRI uses idea management to collect new product ideas and enhancements from a variety of sources. Its customer service staff is always getting ideas from their field experience. The CRI sales force identifies product improvement opportunities based on what customers are telling them that they want, and on what they see CRI's competitors doing. New or emerging technology is an important source of new product opportunities, and the technical leaders at CRI are required to submit new technical ideas every month.

CRI also taps into external sources for new ideas. Customers are an obvious direct source of many good ideas. Using a formal, web-based system permits customers to submit their ideas without seeing the other ideas submitted, and the formal system also enables CRI to rapidly process customers' input and quickly respond back to them. Suppliers have also been a good source of new ideas for CRI, especially ideas for improving the reliability of its products.

Responsibility for idea management at CRI rests with an idea management committee chaired by Susan Kelsey, a strategy manager who heads what CRI calls a market attack team (MAT). This team of five meets monthly to review and categorize all new ideas, solicit additional information where needed, and move new ideas through the new idea pipeline. The MAT participates actively as part of the annual product strategy exercise to ensure that these ideas are considered. Sometimes an idea is so good, or should so clearly be included in a product under development, that Susan's team interjects the idea immediately.

TECHNOLOGY PLANNING

A technology roadmap is somewhat similar to the product line plan in that it is made up of planned technologies, which are similar to planned products. The two differences are that planned technologies do not directly generate revenue and that they are almost always interdependent with planned products.

Planned technologies are timed-phased and contain approximate resource estimates, although these estimates may be less accurate than the estimates for planned products. These resource estimates can be used to plan anticipated resource needs for technology development and help to prioritize technologies on the roadmap to fit within the available resources.

The interdependency of planned technologies with planned products helps to formally guide planning for both products and technologies. When a technology is delayed, the impact of this delay on the products slated to use it can be seen immediately. Likewise, if a planned product is rescheduled, the implications for technologies it uses quickly become apparent.

Figure 20–5 illustrates CRI's technology roadmap and its interdependency with planned products in the Fast-Food Robot product line. Voice-recognition technology is incorporated into the Fast-Food Robot I, but also on the technology roadmap is a plan for second-generation voice recognition that is scheduled to be applied in the Fast-Food Robot II. Third-generation voice-recognition technology is on the roadmap, but its incorporation into specific products has not yet been planned. Other technology projects on the technology roadmap are planned for incorporation into other CRI products, but not the Fast-Food Robot products.

Similar to planned products, CRI uses planned technologies to determine the resource capacity for technology development. As you would expect, the mix of technical skills required is very important to technology development, and this gives CRI as much visibility of shifts in technical skill requirements as possible.

F I G U R E 20–5

CRI's Technology Roadmap

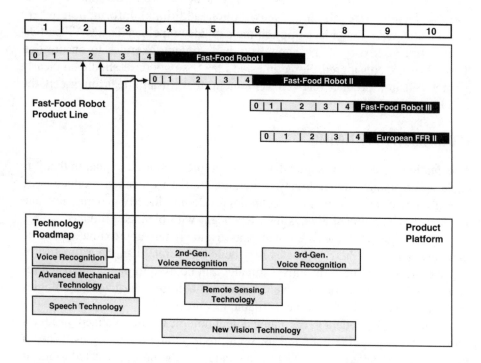

REQUIREMENTS

Integrated product strategy requires a more formal process and a supporting system. In the next generation of product development, the system for product strategy can be an integral part of a DCM system, so that data can be interchanged between product strategy and pipeline management, and alternative product line plans can be simulated against resource capacity. As we have already seen, the integration of product strategy depends on the other systems and processes in the next generation, so it's best to use one highly integrated system.

Product strategy is also a management process, and this process must be formally defined and implemented in conjunction with any new system. The process defines the organizational responsibilities, timing, and approach to developing product strategy. In particular, a company may find it beneficial to separate the executive responsibility for formulating product strategy from the managerial responsibility for overseeing the execution of that strategy. Going back to what was discussed in Chapter 12 on the enhanced phase-review process, the product-approval committee may become more focused on the execution of the product strategy through the phase-review process, and a more strategic group may take formal responsibility for the product line plan.

BENEFITS

There are some important benefits from a formal product strategy process/system that is integrated with product development:

1. Revenue and profit from new products can be optimized by simulating alternative product strategies against development resource constraints.
2. Resource capacity planning can be tailored to anticipated resource needs beyond products already in development.
3. A financial plan for new products can be aligned with the new product pipeline and the product line plan.
4. New product ideas are encouraged from many sources, and some of these may increase new product success.
5. Increased collaboration around product ideas may increase the quality of these ideas, and subsequently the success of new products.
6. The integration of technology plans can better synchronize a technology with the products that use that technology.

SUMMARY

In this chapter we examined product strategy as a system and process in the next generation of product development, with an emphasis on its integration with the broader product development process. Three product strategy processes were explained: product line planning, idea management, and technology planning. In particular, we looked closely at the importance of planned products, how they are created, and how their characteristics are used throughout development.

SECTION

FIVE

Conclusion

21

Getting Started

It's important to recognize that the next generation of product development represents an entirely new way to manage R&D, and not just some new management practices. So implementing this new generation of interlocking processes, practices, and supporting systems will take some time. In fact, it should be looked at more as a journey than as an event. Here again, a look at the history of similar generational changes in major areas in business can provide some guidance. Implementing the MRP Generation in manufacturing took every company years to complete, and the implementation of integrated accounting took as long, if not longer.

Implementing this new generation of product development will also take most companies a couple of years, proceeding one step at a time but steadily building toward a new stage of maturity and performance in development management. Describing this implementation could take up an entire book, so I'm not going to try to get into this in detail. Nevertheless, I can't resist concluding with some broad suggestions for getting started.

DCM SYSTEMS

Throughout this book, I've described how DCM systems now enable a new generation of product development management. I said that DCM is a generic term I use to describe a set of integrated business systems for managing prod-

uct development. This definition was not intended to be a requirements definition for system selection, but merely the minimum description of a generic
DCM system necessary to explain how the new management practices and
processes work. I also stated that the term "DCM" used to describe these systems is not an industry standard term. It may become one, or another may
emerge. Such are the risks of writing about the future.

Companies wanting to move to this new generation have several DCM
systems alternatives. When capabilities such as this are newly emerging, there
are always companies that will invest in creating their own system, but ultimately, as publicly available systems provide better capabilities at a lower
cost, companies abandon their internal systems. In the early days of MRP and
integrated accounting, some companies started to develop their own systems
for these also, but eventually they found that it was too expensive. This will be
the case here as well. A DCM system with sufficient capabilities to do much
of what I've described in this book will cost $10 million to $20 million to develop, possibly a lot more, and the annual cost to maintain and improve it
could be several million dollars each year. Unless a company has genuinely
unique needs, a homegrown DCM system will not make sense.

I hope that I've convinced you by now that software tools are not a DCM
system, but it still might be possible to put together a set of DCM tools in
some way. A set of software tools such as desktop project planning, a resource
database, portfolio spreadsheet, collaboration tool, etc., might, with sufficient
systems-integration work, create a sort of DCM system, but without extensive
integration the same capabilities will not be achieved. Some companies are
currently trying to do this, and may be able to accomplish some of what this
book has described, but they will have to make a lot of compromises along the
way. They will also be hard-pressed to hold their ad-hoc system together over
the long run, but the learning experience could be useful when moving later to
a complete system.

In the long term, the best solution for most is application software for an
integrated DCM system. These applications can provide many of the capabilities
that support the management practices and processes of this new generation.

PROCESSES AND PRACTICES

Throughout this book, I have also made the point that the next generation of
product development is much more than a DCM system. It requires entirely
new management processes, and these processes must be defined and implemented. This won't happen automatically as a by-product of any system. In
some instances, new responsibilities and authorities need to be defined as part

of this new generation; otherwise, there will be confusion over who is supposed to do what with regard to the new processes. There are also a wide range of new management practices that are required to achieve the benefits promised by the next generation. These practices need to be identified, developed, and implemented in order to achieve the benefits I've outlined.

Developing and implementing these new processes and practices is essential to implementing this new generation of product development. Remember that we're not talking about automating the practices of the previous generation, but about wholesale changes to the way product development is managed. So don't make the mistake that many companies made in implementing ERP systems, when they overlooked or minimized the importance of changing the process to take advantage of the new systems.

IMPLEMENTATION PLANNING

After all that I've written, you may have the urge to just wait and see what happens, rather than plunging in. You might well think, "Let other companies do it first," and learn from them. If you wait long enough, your company can hire experienced people from other companies to lead a similar effort for your company. Let the software companies slug it out, see who comes out on top, and then make the safe choice. "Be a follower."

> There are times when it makes sense to be a follower, but this is not one of them. The improvements of the new generation are so profound that they will shift the competitive balance within industries.

There are times when it makes sense to be a follower, but this is not one of them. The improvements of the next generation are so profound that they will shift the competitive balance within industries. Those companies moving to this new generation first will achieve some significant competitive and profit advantages. Many of those who decide to follow will never catch up. As I stated, this is a journey, not an event, so by the time a follower decides to start its journey it may be too late.

There is also what I refer to as the next generation improvement dilemma, which is illustrated in Figure 21–1. Toward the end of a management generation, new tools and techniques promise some incremental improvements. Getting started on a whole new generation, on the other hand, requires some investment to build a new foundation before major improvements follow. At this point, incremental improvements may appear to be better in the

F I G U R E 21–1

The New Generation Improvement Dilemma

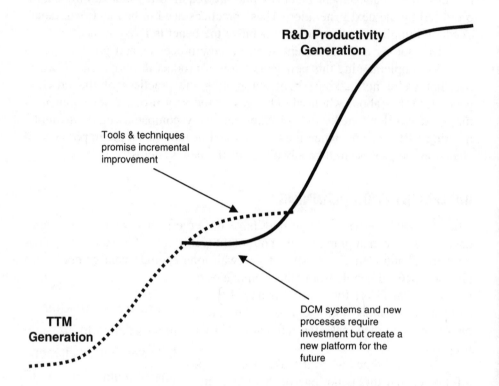

short term. Those familiar with new product platform strategy will recognize this phenomenon. Incremental product improvements can appear to be better in the short term, but new product platforms, a new process generation in this case, are generally preferred.

A lesson learned from the implementation of ERP is not to wait until a long implementation cycle is completed to achieve any benefits, and this lesson can be applied to this new generation of product development. Don't try to do it all at once. Don't even try to plan out a complete, detailed roadmap for everything in the next generation, or you'll spend years planning instead of doing. Instead, I recommend that you learn by doing. Begin implementing some portion of this new generation while keeping a comprehensive vision in sight for the longer term. This comprehensive vision is particularly important when selecting DCM software, because, if possible, you would like to avoid replacing the software later on.

The basic framework of the next generation used throughout this book provides three possible entry points for getting started. You can start through resource management by implementing Stages 1 and 2 of resource management. This will achieve some quick and substantial benefits and give everyone the incentive to continue building this new generation. Implementing Stages 1 and 2 of resource management requires complete resource information from all projects, resource assignments first and then resource needs later. This resource information can be collected initially from all project teams without the need for each of them to implement enterprise project management, but it will quickly provide an incentive for them to do so.

There is also an entry point through enterprise project management. Project teams can begin to use the practices of networked project teams and enterprise project planning and control. This will also provide them with the foundation needed for advanced project management practices. Unlike implementing resource management, implementation here can be done on a pilot basis, with several project teams using the new systems and practices, and this may be a less threatening change in some organizations. Moving on to enhanced, phase-review management will require implementation across all teams.

Finally, portfolio management and product strategy can be used as an entry point. If they are so used, then portfolio and pipeline analysis will still require the manual collection of data, since the project teams would not yet have a DCM system. However, there are still advantages to doing this over using a spreadsheet, since it provides a system that is scalable and able to integrate with future DCM systems, which a spreadsheet cannot do.

CONCLUSION

Well, I didn't promise it would be easy. And it isn't. The next generation of product development is an arduous journey. But as I hope this text and charts and my hypothetical company, Commercial Robotics Inc. (CRI), show, the benefits along the way are worth the journey.

To complete our story of CRI (which my nine-year-old daughter suggested I do), here's how it went for them: The results of using advanced processes and systems were even better than planned. CRI completed the Fast-Food Robot well ahead of its competitors with much more functionality and many more features, leading competitors to think that it must have invested its entire R&D budget on this one project. CRI used the new practices of the R&D Productivity Generation to get much more from its R&D investment than its competitors were able to achieve. The Fast-Food Robot became the dominant product in its market, exceeding all of the original revenue and profit forecasts. The project leader, Anne Miller, was recognized for her ability not only to develop a blockbuster product but also to set an example for the next generation of product development leaders.

Referring to the original benefit model in Figure 2–3, Chris Taylor, CRI's VP of R&D, and Brian Kennedy, its CEO, showed the board of directors how it achieved a 60 percent improvement in R&D productivity through a 30 percent increase in capacity utilization, a 10 percent project efficiency improvement, a 10 percent improvement in developer productivity, and a 10 percent gain from increased pipeline effectiveness. The board was particularly pleased when Shaun Smith, its CFO, explained how the company had achieved financial control and integration of R&D. In sum, it took CRI two years to complete its transition to the next generation of product development; while it wasn't easy being a leader in a new generation of capability, management thought it was well worth the effort and investment.

If there is one thought that I would hope you would take away from reading (or studying) this book—for I realize some of it is quite dense, requiring multiple readings—it is this: We are entering a renaissance of product development capability. The advanced systems and processes available today will join forces with the process capability most companies have grown over the past generation to offer more opportunity and promise for getting better new products to market faster, while doing more with less.

Product development has never been for the faint of heart. Fundamental to this new generation of R&D Productivity is the integration and assimilation of more information by more developers and managers inside the organization. This greater visibility and accessibility will lead to better decision making, better investment in R&D, better return on that investment, more collaboration with external partners, and true empowerment of development activity to meet business goals. There's nothing left to do but . . . get started. I wish you the best.

ENDNOTES

CHAPTER 1

1. *A Guide to the Project Management Book of Knowledge* (Project Management Institute, 2000), p. 200.
2. Ibid., p. 204.
3. For a complete description of PACE®, see *Setting the Pace in Product Development: A Guide to Product and Cycle-time Excellence,* ed. by Michael E. McGrath (Butterworth-Heinemann, 1996).
4. The term "stage-gate" was coined by Robert G. Cooper and is a trademark of RG Cooper and Associates.
5. See "The R&D Effectiveness Index," by Beth Ginsberg and Michael McGrath, in *PRTM's Insight,* vol. 10, no. 2 (Summer 1998), pp. 14–17.

CHAPTER 6

1. PRTM research on hidden costs in R&D (paper dated March 23, 2001).

CHAPTER 14

1. *A Guide to the Project Management Book of Knowledge,* p. 204.

CHAPTER 17

1. Michael E. McGrath, *Product Strategy for High-Technology Companies* (McGraw-Hill, 2001).

CHAPTER 18

1. For example, see Mark Deck, "Beyond Projects: Creating a Winning Product Portfolio," in *PRTM's Insight,* vol. 15, no. 1 (Fall 2003), pp. 17–23.

CHAPTER 20

1. *Product Strategy for High-Technology Companies.*
2. Shira P. White, with G. Patton Wright, *New Ideas About New Ideas: Insights on Creativity from the World's Leading Innovators* (Perseus Publishing, 2002).
3. Ibid., p. 178.

GLOSSARY

This book introduces some terms that may be new to some readers, and uses other terms in a precise way. In order to help readers understand these terms, they are defined below.

Assignments, project—Resource assignments of specific individuals to a project for a defined period of time.

Bottom-up project planning—Project planning, where project activities are defined at the detailed task or activity level, the time required for each task or activity is estimated, and then the overall project schedule is estimated by "rolling up" task and activity times.

Capacity management, R&D—Management of all R&D resources in order to maximize the utilization of these resources and avoid bottlenecks.

Closed-loop time collection—System for comparing actual time charged to projects by developers to their scheduled time.

Collaboration system services—A model for collaboration software that defines the layers of collaboration capabilities for different types of collaboration software.

Collaborative product development—Developing products across corporate boundaries in conjunction with external partners who contribute to the development of the product.

Context-based knowledge management—The capability of delivering relevant knowledge and experience to the individual developer when he or she needs it.

Cross-functional core team—Hub-and-spoke project team model, with small inner team empowered to coordinate all project activities of the extended team. Inner team members are drawn from different functions, with a project manager facilitating the team.

Development chain management (DCM) systems—Integrated systems that automate product development management, including systems for managing processes, knowledge, and data used in resource management, project management, and portfolio management/product strategy.

Distributed program management—The capability to distribute the management of related projects while simultaneously managing these as an integrated program.

Enterprise project planning and control—A collaborative or shared process and system under which each member of the project team collaborates to develop a common plan and uses this common plan to control all steps in the project.

Financial planning for new products—The projection of financial results from a new product.

Idea management—The process by which a company collects, refines, archives, and prioritizes ideas for new products or product improvements.

Invention and Commercialization Generation—The initial generation of product development management, starting approximately 1880.

Long-term resource planning—Resource planning, typically at an aggregate level, for a time horizon of a year or more.

Medium-term resource planning—Resource planning based on anticipated resource needs in addition to current project assignments, typically for a planning horizon of 3 to 18 months.

Needs, project resource—Anticipated project resource requirements defined by a skill or skill set for a period of time, not an individual.

Networked project teams—Project team model that enables all team members to coordinate and communicate through a virtual project network system, using coordinating mechanisms such as project schedule elements, calendars, etc.

Outsourcing, R&D—Moving R&D activities from internal resources to external resources, usually with the intent to lower costs.

PACE® (Project and Cycle-time Excellence®)—A framework developed by PRTM for the new management practices in the Time-to-Market Generation, including phase-review management, cross-functional core teams, structured development, and portfolio management, platform strategy, product line strategy.

Phase-review process—Decision process where senior management (a Product Approval Committee) approves new product development project investments on a phase-by-phase basis.

Pipeline management—The way resources are deployed and investments are prioritized to ensure that the projects chosen for investment have the resources they need to be successful.

Planned product—A product that is not yet in development, but is formally on the product line plan to be developed. Planned products are the essential elements of product strategy that contain its characteristics, such as resource estimates, financial projections, etc.

Portfolio management—The way a company selects and prioritizes a group of projects to achieve its business goals, especially maximizing the long-term value of new product investments.

Product Approval Committee (PAC)—The cross-functional senior management group that has the authority and responsibility to approve and prioritize new product development within a phase-review process.

Product line plan—A time-phased conditional plan for sequencing development of new products in a product line that is the result of the product strategy process.

Program—A set of related projects such as product variations, nonproduct-development projects related to a product being developed, multiple products within a common product line, or subprojects within a broader project.

Project budget—The expected investment of resources, expenses, and capital to develop a new product.

Project Success Generation—The generation of product development management from approximately 1950 to the mid-1980s, which focused on making individual projects successful.

PRTM—Pittiglio Rabin Todd & McGrath, a global management consulting firm (www.prtm.com) founded in 1976.

R&D Effectiveness Index—A quantitative measure of R&D performance introduced by PRTM in 1992—and benchmarked six times during the 1990s—that measures the profit from new products compared to R&D investment.

R&D productivity—The output (new product profit) of R&D investment, compared to that investment.

R&D Productivity Generation—The generation of new product development starting at the turn of the 21st century, which focuses on a new approach to management of development based on the enabling technology of development chain management systems.

Resource capacity—The total time of all developers prior to assignment to development projects or other activities.

Resource charge-out—The process for charging resources from resource groups to projects, usually with the objective to fully allocate all the costs from a resource group.

Resource costs—The total cost per unit of time (month, day, hour, etc.) of a person assigned to a project. Various methods are used to determine these costs.

Resource group—A department or function, usually within R&D, but sometimes outside of R&D, that has responsibility for managing a group of people who are assigned to product development projects.

Resource management—The management of a company's investment in development resources, primarily its people, in order to achieve the optimum product development output consistent with its strategy.

Resource requirements management—The process of effectively allocating assigned developers to specific project steps.

Resource requirements planning (RRP)—The process of determining resource needs for a project with the objective of achieving the most efficient use of resources.

Resource transaction process—The process that controls the initiation of resource requests for approved projects from needs identified by project managers, the response to these requests with assignments by resource managers, and the coordination of all exceptions, changes, and adjustments to these resource transactions.

Self-organizing—The ability of members of a project team to work together dynamically using the coordinating mechanisms of a project network system.

Short-term resource planning—Resource planning based on project assignments with a typical planning horizon of one to three months.

Structured development process—Defines the hierarchy of project phases, steps, and tasks, enabling companies to apply top-down cycle-time standards to project planning.

Technology roadmap—A time-phased plan for new technologies related to the planned products using that technology.

Time to market (TTM)—The product development cycle time from initial product concept to successful market entry.

Time-to-Market Generation—The generation of new product development management that started in the mid-1980s and continued into the early 2000s, which introduced new management practices primarily aimed at reducing time to market.

Top-down project planning—Project planning that starts at the highest level (the project approval or phase level), with schedule-items expanded to the next level in more detail.

Utilization, R&D capacity—Percentage of R&D resources assigned to approved revenue-producing projects. This could be at a resource group level or overall.

Work management—Management of project activities at the lowest level of detail for the purpose of daily coordination.

INDEX

Accountability, issue, 61
Accounting
    integration, 26–27, 354
    systems, 335
        integration, 27
Accounts payable department, 27
Accounts receivables
    department, 27
    invoices, 26
Action-items, 261, 273
    management, 195
Actual costs
    inclusion. *See* Project budgets
    management, example, 236–238
Actual-to-budget comparison, 237
Advanced DCM system, usage, 221
Airplane, commercialization, 7
Alternatives, 177
    identification, 176
Annual financial planning
    medium-term resource capacity
        planning integration, 104–105
    resource management, integration,
        142–143
Applied developer, real cost, 52
Approvals, vertical organization, 182
ASP (average selling price), 225
Assembly-line mentality, 181
Assignment management. *See* Formal
        assignment management
    system, impact, 87

Assignment-based utilization, 60
Assignments
    charge out, 142
    definition. *See* Project assignments
    derivation. *See* Project budgets
    distribution, 132–133
    format, 95
    information, usage, 91
    planning, 85, 110
    request, procedure, 95
    translation, 138
Automated financial planning, 241
Automated integration, usage, 148
Automated resource transaction process, 117
Average selling price (ASP), 225

Baseline performance, 82
Bell Laboratories, 8
Best-practice standard models, development,
    282
Best-practices guidelines, 159
Bill-of-material (BOM), 227
    applications, 218
Binney, Edwin, 8
Binney & Smith, 8
Board of Longitude, 6
BOM. *See* Bill-of-material
Booz, Allen, and Hamilton, 9
Bottom-up architecture, 166
    contrast, 166–167

Bottom-up process, integration, 74
Bottom-up project planning, definition, 361
Bottom-up work estimates, 124
Bubble economy, 17
Budgets. *See* Project budgets
    example. *See* Capital budgets; Expense
    budgets
Bulletin boards, project, 196
Bundy Manufacturing Company, 7
Business-level revenue projection, 327

CAD/CAE software
    integration, 218
    usage, 190
CAD/CAE systems, usage, 286
Calendar items, 261, 273
    maintenance, 253
Capacity
    management, 57. *See also* Medium-term
        capacity management; Pipeline capac-
        ity management; Research & devel-
        opment capacity
        process, 66
    planning. *See* Long-term resource capacity
        planning; Medium-term resource ca-
        pacity planning; Short-term resource
        capacity planning
        skill, 109, 306
    utilization, 50
        improvement. *See* Research & develop-
            ment capacity
        increase, 39
Capital budgets, 219
    example, 233
Capital expenses, 233
Capital plan, consolidation, 330
CFO
    capabilities, 158
    control, 335–336
    memo, 238
    responsibility, 323
    role, 300, 307. *See also* Research & devel-
        opment
Chain libraries. *See* Development chain man-
    agement system
Chain-of-command functional team structure, 183
Channel partners, 267

Charge-out approach/process, usage, 325, 332
Charting techniques, 314
Church, Derry, 8
Clark, Kim, 13
Closed-loop time collection, 80, 146–148
    definition, 361
Codevelopers, usage, 145, 159
Collaboration, 33. *See also* Shared documents
    e-mail basis, 268–269
    enabling, 272
    strategy, 275
    systems services, 267–273
        definition, 361
    web-based workspace, usage, 269–270
    workspaces, 269–270
Collaborative development, 155
    benefits, 37–38
    categories, 260–267
    management, 159, 259
        benefits, 275–276
        requirements, 274–275
    strategy, 266
Collaborative document preparation, 198–199
Collaborative product development, defini-
    tion. *See* Product development
Collaborative project control, 169, 172–173
Collaborative project planning, 169–172
Colocation, 185
    environment, 188
Combined phase reviews, 209
Command-and-control hierarchical structure, 14
Commercialization
    airplane, 7
    competency, 6
Commercialization Generation, 5–9
    definition, 362
Communications, 202
    function, distribution, 187
    gratuitous communication, 186
    problems, organizational solutions,
        184–185
    structure/coordination, 187–190
Competitive products, evaluation, 193
Complex project management, 247
Component libraries, 287
Comprehensive financial management. *See*
    Research & development, comprehensive
    financial management

Computing-Tabulating-Recording Company (C-T-R), 7
Consolidated program view, providing, 159
Content management, networked project teams, 197–199
Context-based knowledge management, 155, 159–160, 277
  benefits, 290–291
  definition, 361
  requirements, 290
Contract manufacturing, 274
Contractors, integration, 265–266
Cooper, Robert G., 13
Coordinating mechanisms, 191
Core-functional core team, improvement, 32
Core-team best practices, 183
Core teams, project, 201
  cross-functional management, core team, 182–188
  distributed e-mail communications, 185–188
Corporate development organization, 183
Cost estimates, example. *See* Product cost estimate
Cost layers, 92
Cost-reduction opportunities, 37
Cost savings, recovery, 240
CPM. *See* Critical Path Method
Crayola crayons, 8
Critical-chain resource management, 131
Critical-chain techniques, 133
Critical Path Method (CPM), 9–10
  usage, DuPont, 9
Cross-functional core team, 154
  creation, 156
  definition, 361
  management, 182–188
  model. *See* Project And Cycle-time Excellence
    optimization, 183
  organization, 185
  traditional model, 183–185
Cross-functional process, 182
Cross-functional project core team, 204
Cross-functional project teams
  introduction, 12
  organizational model. *See* Time to market Generation

Cross-functional teams, 14–15
Cross-project interdependencies, 250
Cross-project level, resource management, 72
Cross-project management, 15
C-T-R (Computing-Tabulating-Recording Company), 7
Customers, 261–264
  deliveries, reliability, 26
  feedback, evaluation, 193
  involvement, 262
  meeting, notice, 196
  service, 274
Customer/vendor approach, 10
Cutting-edge practices, 4
Cycle-time standards, usage problems, 205
Cycle times, planning standards, 165

Data, manual collection, 311
Dayton Engineering Laboratories, 8
DCM. *See* Development chain management
Decision making. *See* Phase-based decision making
  importance, 326
  process. *See* Product development
Deliverables, project, 194
Design documents, maintenance, 190
Design review meeting location, change, 196
Developer productivity, 36–37
Development
  activities, 15
  cost, 216
  cycle, 10
  management. *See* Collaborative development, management
  partners, 264–265
    external, 263–264
  product
    processes, standard, 15–16
    definition. *See* Structured development process
    programs. *See* Product development projects
      integration. *See* Technology organization portfolio, 164–165
  teams, work coordination, 257

Development chain management (DCM)
    capability, 272
    enterprise planning/control system, 159, 271
    extension, 330
    partnering capabilities, usage, 265, 270
    processes/practices, 354–355
    product development knowledge, delivery,
        286–289
Development chain management (DCM) sys-
        tem, 21, 41, 116, 141
    automated tool, usage, 123
    chain libraries, 173
    definition, 361–362
    enterprise project planning/control system,
        169
    information technology, usage, 156
    initiation, 353–354
    integration, 43, 84, 136, 272–273. *See also*
        Enterprise resource planning systems
    partnering
        capabilities, integration, 270–272
        extension, 260
    single integration, 149
    usage, 30, 113–114, 155, 162. *See also*
        Advanced DCM system
Development resources
    costs
        budgeting, 231
        example, 230–231
    waste, reduction, 17
Dey, Alexander, 7
Distributed e-mail communications, core
        teams, 185–188
Distributed program management, 155,
        158–159, 243
    benefits, 257
    definition, 362
    focus, 248
    requirements, 256–257
Distributed program planning/control,
        249–250
Distributed project communications, 189
Distributed project team coordination,
        250–253
Documents. *See also* Reference documents
    backup/archiving, management, 198
    collaboration, 251
    collaborative preparation, 198–199

    project, management
        capabilities, 199
        read-only format, 197
        usage, 261, 273
DuPont, CPM (usage), 9
Duration-analysis technique, 9
Dynamic pipeline management, 305–306,
        309, 315–321
    benefits, 321
    objective, 318
    requirements, 321
Dynamic portfolio management, 305–306,
        309, 311–315
    benefits, 321
    requirements, 321

Earned value analysis (EVA), 236
    example, 238–239
Edison, Thomas Alva, 6
Educational materials, 284–285
Effectiveness Index, definition. *See* Research
        & development
E-mail-based applications, disadvantages, 200
E-mail basis, collaboration, 268–269
E-mail communications, distributed to core
        teams, 185–188
E-mail services, 268
E-mail usage, 189
Endnotes, 359–360
Enhanced phase-review management, 203
    benefits, 213
    improvements, 207–213
    requirements, 213
Enhanced phase-review process, 157
Enterprise project control, 155–156
    definition, 362
    example, 175–178
    integration. *See* Resource management,
        integration
    process, 161, 169–175
        benefits, 178–179
    system, 162, 289
    usage, 249
Enterprise project management, 155–157
Enterprise project planning, 156
    architecture, 163–169
    benefits, 178–179

Enterprise project planning (*Cont.*)
   definition, 362
   example, 175–178
   integration. *See* Project planning
   process, 161, 169–175
   system, 162, 289
   usage, 249
Enterprise resource planning (ERP) systems,
    144, 220, 236
   arrival, 325
   criticism, 163
   DCM system integration, 222
   implementation, 356
Enterprise system
   integration, 313
   project planning tool, contrast, 164
Enterprise-wide information systems, 325
ERP. *See* Enterprise resource planning
Estimates, weighted average, 10
EVA. *See* Earned value analysis
Event-oriented network analysis technique, 10
Expected resource requirements, usage. *See*
  Functional budgets
Expense budgets, 232
   example, 231
   usage, 219
Experience, types, 278–285
External collaboration, enabling, 272
External development partner, 263–264
External information, 285
External resources, resource management (in-
  tegration), 144–146

FAQs (frequently asked questions)
   list, 283
   review, 289
Feedback
   collection. *See* Quick status feedback
   evaluation. *See* Customers
File server, usage, 269
Financial analysis, 67, 234
   example, 233–235
Financial conference, president speech, 196
Financial data, 261
   integration, 241
   real-time consolidation, 300
   source data, impact, 221–222

Financial executives, 296
Financial information, 323
   consolidation, 158, 223–224
   embedding, 235
   incorporation. *See* Project documents
   integration, 217–224
   product financial information, 326–331
Financial management
   application, 239
   integration. *See* Program
   process level, 299–300
   product development, 217
   product project, 236–240
   research development, 36, 39
Financial models, 281, 282
Financial plan, 334
   long-term, 328
   product, example, 229–230
Financial planning, 32. *See also* Annual
    financial planning; Automated financial
    planning
   definition, 362
   inconsistency, 224
   integration, 155, 215, 222
    example, 224–230
   models, implementation, 158
   project budgeting, integration, 158
Financial returns, projections, 216
First-to-market advantage, 12
Fixed asset purchases, 26
Fixed-capacity operation, 50
Fixed R&D capacity, output maximization, 62
Ford, Henry, 7
Ford Motor Company, 7
Formal assignment management, 88
Frequently asked questions (FAQs)
   list, 283
   review, 289
Full-time equivalents (FTEs), 72–73
   calculation, 74, 109, 130
   need, 112
   person-day transformation, 133
   reduction, 131
   usage, 129
Fully integrated resource management, 139
   benefits, 149
   requirements, 149
Functional budgets, 324

Functional budgets (*Cont.*)
  allocation, tracking, 332–334
  expected resource requirements, usage, 334
  project budgets, reconciliation, 331–334
  resource management, reconciliation, 142
Functional design, preliminary development, 193
Functional project organization, variation, 10–11

Gantt, Henry, 9
Gantt chart, creation, 9
General Motors, 8–9
  Research Corporation, 8
General Technical Committee, 8
Genius, definition, 6
Gratuitous communication, 186
Gross margin tolerance, 209
Groups, management. *See* Resource groups
Gulf War, 27–28

Harrison, William, 6
Hershey, Milton, 8
Hershey Chocolate Company, 8
High-level design, consensus, 177
High-level pricing task, 288
High-level resource profile, 342
High-level tasks, 192, 287
  management, 175–176
Hollerith, Herman, 7
HR (human resources) systems skill categories, resource management (integration), 144
Hub-and-spoke arrangement/design, 183–184
Hub-and-spoke organization, 15, 185
Human resources (HR) systems skill categories, resource management (integration), 144

IBM, incorporation, 7
Idea management, 344–347
  definition, 362
Implementation planning, 355–357
Industrial Revolution, 5–6
Industry benchmarks, 16

Industry journal article, usage, 196
Information
  external, 285
  integration, 25, 217–224
  system. *See* Enterprise-wide information systems; Product approval committee
  technology, 274
  transferring, 141
  transformation, 24–28
  usage, assignments, 91
In-house system-test department, usage, 67
Initial stage improvements, 59
Institutional experience, 282
Integrated accounting. *See* Accounting
Integrated financial planning. *See* Financial planning, integration
Integrated pipeline management process, 320–321
Integrated portfolio management process, 314–315
Integrated product strategy. *See* Product strategy
Integrated project budgeting. *See* Project budgeting
Integration
  philosophy, 140
  strategy, 146
  usage, automated, 148
Internal correspondence, usage, 273
Internal rate of return (IRR), 235
Invention Generation, 5–9
  definition, 362
Inventory, 26
  levels, 24
  re-order point, 24
IRR (internal rate of return), 235
Issue-resolution items, 195–196

Just-in-time training, 285

Kettering, Charles, 8
Key-word search engines, 278
Knowledge
  collection, product development, 289
  completeness, 28
  librarian, 289

Knowledge (*Cont.*)
  linking, 287–289
  types, research & development, 278–285
Knowledge management. *See also* Context-based knowledge management systems, 277
    resource management, integration, 148

Lancaster Caramel Company, 8
Libraries, step, 286–287
Longer-term capacity planning, 343
Long-term financial plan, 328
Long-term resource capacity planning, 106
  aggregate level implementation, 106
Long-term resource estimates, 101
Long-term resource planning, 100
  definition, 362

Management
  capability, requirements, 94
  process, 348
    architecture, 163
    levels, 43–44
Market
  coverage, 309
  expectations, analysis, 193
  growth estimates, 329
  opportunities, 38–39
  study requirements, shifting, 129
Materials requirements planning (MRP), 24–26
  generation, 353
  systems, implementation, 25–26
Matrix organization, 11
Matrix teams, 182
Maturity, stage, 75, 155
Medium-term capacity management, 99
Medium-term capacity resource management, visibility, 78
Medium-term resource capacity management, 78
Medium-term resource capacity planning, 78, 99, 103–104
  benefits, 117
  improvement, 143
  integration, annual financial planning, 142–143
  requirements, 115–117

Medium-term resource planning
  definition, 362
  extension, 105, 117
MRP. *See* Materials requirements planning
Multiple resource groups, resource management (integration), 144

Need to know position, 268
Net present value (NPV), 220
Network analysis technique, event-oriented, 10
Networked project teams, 156, 181, 188–189
  benefits, 201–202
  content management, 197–199
  coordinating mechanisms, 201
  definition, 362
  experience, 191
  project communication, 189–191
  project coordination, 191–196
  requirements, 200–201
  self-organization, 193–195
Networked teams
  network, 199–200
  project network system, usage, 189
  project plan, usage, 175
Next generation
  improvement dilemma, 355–356
  promise, 3
Non-value-added project activities, 36
NPV (net present value), 220

OEM, enabling, 266
On-demand portfolio analysis, 313–314
Operation Desert Storm, 28
Operation Iraqi Freedom, 27–28
Organizational filtering/distortion, 184
Organizational responsibilities, 348
Outsourcing. *See also* Research & development outsourcing
  definition. *See* Research & development
  gains, partnering/outsourcing, 37–38
  increase, 274
Overscheduling, 90

PAC. *See* Product Approval Committee
Pace, General Peter, 27–28

PACE. *See* Project And Cycle-time Excellence
Partnering capabilities, 259
  integration. *See* Development chain management system, inegration
Partnering/outsourcing, gains, 37–38
Partners. *See* Channel partners; Development
Payroll processing, 26
PDM, usage, 218
Performance, baseline, 82
  evaluation, 90
Periodic pipeline management, 303–305, 316
Periodic portfolio management, 303–305
Periodic portfolio rationalization, 310
Person-day
  estimates, 133
  transformation. *See* Full-time equivalents
PERT (Program Evaluation and Review Technique, 9–10
Phase-based decision making, 12–14, 157, 203
  limitations, 206
Phase-based decision process, 204
  requirement, 13, 157
Phase-review information system, 213
Phase-review management process, 204
Phase review process, 13, 85, 204–205
  characteristics, 205
  definition, 362
  enhanced, 203
  implementation, 212–213
  limitations, 206
  management, 211
  usage, 94, 141
Phase-review product-approval process, 314
Phase-review project-approval process, usage, 103
Phase reviews, 203–206
  combined, 209
  decisions
    impact, 102
    resource availability information, usage, 207–209
  focus, 212–213
  management, 32
  presentations, 197
Physical inventory, 26

Pipeline analysis, 29
Pipeline capacity management, 105–106
Pipeline management, 16–17, 105–106. *See also* Dynamic pipeline management; Periodic pipeline management
  absence, 303
  benefits, 38
  definition, 362
  process
    integration, 320–321
    level, 300–301
  resource management, integration, 141
Pipeline simulation/optimization, 318–320
Pittiglio Rabin Todd & McGrath (PRTM), 12
  definition, 362
  PACE implementation, 167
  study, 113
  usage, 18
Placeholder, creation, 66
Planned assignments, 85, 110
Planned products, 338, 341–343
  definition, 362–363
  financial projections, 299
  standard profiles, 343–344
Planned project, launch date, 346
Planning. *See* Annual financial planning; Project planning
  models, 220
  standards, 170
Plan-of-record, 309
Platform line strategy, 339
PLM (product lifecycle management), 41
Portfolio
  alternative, 318–320
  analysis, 29, 313–314
  simulation, 30
Portfolio-level management, 296
Portfolio management, 16–17, 29–30, 357
  See also Dynamic portfolio management; Periodic portfolio management
  absence, 303
  definition, 363
  limitation, 310
  process
    integration, 314–315
    levels, 297–302
  resource management, integration, 141
  stages, 295, 302–308

Practices, information transformation, 24–28
Preliminary functional design, development, 193
Pricing strategies, 278
Probabilistic-weighted revenue/profit forecasts, 320
Processes
    development, coordination. *See* Product development
    information transformation, 24–28
    innovation, 7
    levels. *See* Financial management; Management; Pipeline management; Portfolio management; Product strategy integration, 301–302
Product Approval Committee (PAC), 13–14, 112, 157, 204–209
    approval, 235
    bulletin board, 212
    definition, 363
    information system, 211–212
    meeting, 315
    review, 64
Product cost estimate, example, 227–229
Product design change, 221
Product development, 3, 24, 54
    collaboration, definition, 361
    decision-making process, 12
    financial management, 217
    generation, adoption, 4–5
    knowledge
        collection, 289
        delivery. *See* Development chain management
    management
        generation, 4–5
        history, 4
    output, 117
    process, structured, 173
        development, coordination, 248–249
    programs, 245–249
    projects, 58, 245
        integration. *See* Technology
    TTM Generation, 299
Product expense, consolidation, 331
Product financial analysis, 219–220
Product financial information, consolidation, 326–331

Product financial plan
    creation, 218–219
    example, 229–230
Product gross margin, 210
Product lifecycle management (PLM), 41
Product lifecycle simulation, 30
Product line plan, 340–344
    definition, 363
    feasibility, simulation, 344
Product line strategy, 339
Product planning financial model, 288
Product platform lifecycles, 298
Product profitability, consolidation, 331
Product project, financial management, 236–240
Product strategy, 29–30, 337
    integration, 307–308
    process, 338–339
        levels, 297–302
    resource management, integration, 143–144
    stages, 295, 302–308
    system, 338–339
Product strategy process, integration, 337
    benefits, 349
    requirements, 348–349
Production efficiencies, 7
Productivity benefits, 30
Productivity Generation
    benefit model, 35–41
    definition. *See* Research & development Productivity Generation
Products. *See also* Planned products
    approval process, 296
    ideas, management, 30
    investments, long-term value, 309
    lifecycle, 340
    multiple versions, coordination, 246
    profiles. *See* Planned products
    requirements, 25
    revenue forecasts, 307
Profit
    adjustment, 62
    forecasts, 297
        probabilistic-weighted revenue, 320
Program
    budgeting, 246
    coordinators, 182

Program (*Cont.*)
  costs, 255–256
  definition, 363
  financial management, integration,
    253–256
  management, 243. *See also* Distributed
    program management
  planning/control, distributed, 249–250
  revenue, 254
Program Evaluation and Review Technique
  (PERT), 9–10
Program-level collaboration, shared, 251
Progress, reporting, 172–173
Project action, items, 194–195
Project activities, non-value-added, 36
Project And Cycle-time Excellence (PACE)
  cross-function core team model, 14
  definition, 362
  implementation. *See* Pittiglio Rabin
    Todd & McGrath
  process, 12, 154
    example, 167
  structured development process, 15
  terminology, 157
  usage, 76
Project approval process, 85
Project assignments
  definition, 361
  making, 95
Project-based management, 256
Project budgeting, 32
  definition, 215–216
  integration, 215. *See also* Financial
    planning, integration
    example, 230–235
Project budgets, 210, 219, 324
  alignment, 221
  consolidation, 331
  definition, 363
  inaccuracy, 224
  reconciliation, 331–334
  resource assignment derivation, 220–221
  resource management, integration, 143
  updates, actual costs (inclusion), 222
Project bulletin boards, 196
Project calendar, items, 193–194
Project communication, networked
  project teams, 189–191

Project contract system, development, 13
Project control. *See* Collaborative project
  control; Enterprise project control
Project core team, 201
Project data, real-time integration, 311–313
Project deliverables, 194
Project documents
  financial information, incorporation, 222
  management, 197–198
Project financial analysis, example, 233–235
Project-level productivity, improvement, 70
Project management, 32–33, 148
  advanced practices, 158–160
  complex, 247
  disciplines, 154
  enterprises, 155–157
  internal market form, 11
  stages, 153
  step level, 166
  system, 43
Project managers, 112
  external resources, usage, 145
  resource discovery, 69
  resource requests, 63
  software tools, usage, 161
Project planning. *See also* Enterprise project
    planning
  bottom-up, definition, 361
  collaborative, 169–172
  framework, 168
  integration. *See* Resource management
  resource management, integration, 146
  resource needs, impact, 107–109
  standards, 179
    enterprise project planning/control, in-
      tegration, 173–174
  tool, contrast. *See* Enterprise system
  top-down, 15
  work management, contrast, 163,
    167–168
Project plans, 281, 284
  enterprise-wide system, 162
  hierarchical structure, 164–166
  integration, 174–175
  sharing, 271
  synchronization, 247
  usage, networked teams, 175
Project requests, acceptance, 103

Project resource
  efficiency, increase, 35–36
  management, real-time integration, 316–317
  needs, 66, 106–112
    definition, 362
Project resource requirements
  optimization, 73
Project resource requirements management
    (RRM), 119
  benefits, 137–138
  requirements, 136–137
Project resource requirements planning
    (RRP), 119–129
  benefits, 137–138
  requirements, 136–137
Project schedule items, 192–193
  interdependency, 169
Project Success Generation, 9–12, 303
  definition, 363
Project teams, 158. *See also* Networked pro-
    ject teams
  coordination. *See* Distributed project team
    coordination
  dynamic networking, 200
  managers, 83
  members, 189
    introduction, 196
  organization, 10–12
  self-organization, 175
Project tolerances, 209–210
  automatic monitoring, 213
  criteria, 210
Projects
  approval process, 55
  completion dates, 155
  coordination, 244. *See also* Networked
    project teams
  financial return, 28
  functional budget allocation, tracking,
    332–334
  information, 340
  ROI, 79, 119
  scheduling, 9–10
  steps, 192
    distributed responsibility, 179
    time phasing, 126–127
  three-level architecture, 165–166
  variations, 306

PRTM. *See* Pittiglio Rabin Todd & McGrath
Punch Card Tabulating Machine, 7

Quasi-autonomous transactions, 10
Quick status feedback, collection, 148

R&D. *See* Research & development
RDEI. *See* Research & Development Effectiveness Index
Real-time actual project data, 29
Real-time data, 305
Real-time images, 27
Real-time integration. *See* Project data; Project resource
Reference documents, 283–284
Research & development (R&D)
  budgets, 143, 324
  CFO, role, 324–326
  comprehensive financial management,
    306–307, 323
    benefits, 335–336
    requirements, 335
  cost, 93
  effectiveness, increase, 22
  executives, accountability, 53–54
  experience
    gaining, 282–283
    types, 278–285
  financial management, 30, 39
  guidelines, 279–281
  investment, 275
    deduction, 34
    distribution, 304
  knowledge, types, 278–285
  maintenance, 274
  management, 5
    next generation expansion, 49
    TTM Generation, 153
  outsourcing, 53, 274
    definition, 362
  policies, 279–281
  productivity, 50
  resources, 103
    demand/supply balancing process, 72
    utilization, 54
  spending, 334

Research & development (R&D) (*Cont.*)
    standards, 279–281
    templates, 281–282
    utilization, 75
        increase, 62
        understanding, 55–57
Research & development (R&D) capacity
    increase, 40–41
    management, 361
    measurement, 59
    output maximization, fixed, 62
    portfolio alternatives, 106
    utilization
        definition, 364
        improvement, 35
Research & Development Effectiveness Index
    (RDEI), 18
    definition, 363
Research & development (R&D) Productivity
    Generation, 23, 45
    characteristics, 28–33
    definition, 363
    genesis, 20–22
Resource assignments, 115
    derivation, 220–221
    linking, 96
    management, 84, 96
    requesting, 145
    short-term management, opportunity, 31
    step-level, 143
Resource availability, 76, 85, 126, 168
    confirmation, 63
    information, usage, 103
    visibility, increase, 117
Resource-based phase-review decisions, 213
Resource-based project planning, 162
    requirement, 174
    usage, 168
Resource buffers, usage, 138
Resource capacity
    definition, 363
    planning. *See* Long-term resource capacity
        planning; Medium-term resource capac-
        ity planning; Short-term resource capac-
        ity planning
    visibility, 110
Resource charge-out, definition, 363
Resource conflict, resolution, 128–129

Resource cost, 93
    definition, 363
    determination process, 95
    example. *See* Development
Resource demand
    horizon, 317
    visibility, 110
Resource estimates, 284
    balancing, 138
    making, 123–126
Resource groups
    definition, 363
    headcount plans, 142
    management, resource needs (impact),
        110–112
    resource management, integration. *See*
        Multiple resource groups
Resource information, definition, 71
Resource management, 30–31. *See also* Criti-
    cal-chain resource management; Fully
    integrated resource management
    benefits, 76
    by-products, 60
    capability, 88
    cross-project level, 72
    definition, 363
    focus, shift, 31
    improvement/benefits, exten-
        sion/institutionalization, 79–80
    insufficiency, 70
    integration, 140. *See also* Annual financial
        planning; External resources; Knowl-
        edge management; Pipeline manage-
        ment; Portfolio management; Product
        strategy; Project budgets
        points, 141–148
    levels, 71
    project planning, integration, 162,
        168–169
    reconciliation. *See* Functional budgets
    three-level framework, 80–81
Resource managers, 76
    demands, anticipation, 65
Resource needs, 113. *See also* Project re-
    source; Step-specific resource needs
    consolidation, visibility increase, 117
    impact. *See* Project planning; Resource
        groups management

Resource needs (*Cont.*)
  resource requirements, transformation, 129–131
  skill categories, 109–110
Resource planning
  definition. *See* Long-term resource planning; Medium-term resource planning; Short-term resource planning; Top-down resource planning
  time horizons, 100–101
Resource profile, high-level, 342
Resource requests, 113–115
Resource requirements
  balancing, 128–129
  estimation, 123
  expected usage, 334
  guidelines, usage, 123
  management, 73, 131–133
    definition, 364
  transformation, resource needs, 129–131
  translation
    ability, 120
    tools, 137
Resource requirements planning (RRP), 41, 73. *See also* Project resource requirements planning
  definition, 364
  techniques, usage, 130–131
  tools, 136–137
Resource review meetings, 209
Resource transaction process, 112–115. *See also* Automated resource transaction process
  definition, 364
  impact, 113
Resource transaction-management process, 116
Resource utilization. *See* Research & development
Resources
  allocation, 207, 332
  balancing, 138
    time-phasing steps, 126–127
    tools, 137
  charge-out, 66
  charging, 142
  constraint, 304
  demand, 72

Return on investment (ROI), 220, 235. *See also* Projects, ROI
  weighted, 331
Revenue forecasts, 226, 297, 343. *See also* Probabilistic-weighted revenue/profit forecasts; Products
  non-reconciliation, 224
Revenue-generating projects, 58
Revenue projection, 62, 223, 230
  business-level, 327
  consolidation, 327–330
  example, 224–227
Revision control, 198
  example, 239–240
ROI. *See* Return on investment
Roles, knowledge (linking), 287–289
Rolled-up item, 171
Rolling up, appropriateness, 166
RRM, project. *See* Project resource requirements management
RRP. *See* Resource requirements planning

Schedule item, 188, 261, 273. *See also* Project schedule items
  definition process, 172
  dependencies, 250
  interdependencies, 170
Scheduling. *See* Overscheduling; Projects; Tasks
Search engines, key-word search engines, 278
Second Stage Improvements, 59
Self-organization, 202. *See also* Project teams
Self-organizing, definition, 364
Senior development managers, 295
Server-based collaboration, 269
Shared documents, program-level collaboration, 251
Shared file server, usage, 269
Short-term resource capacity planning, 102–103
Short-term resource planning, 100
  definition, 364
Skill categories, 109
  HR systems, resource management, 144
  resource needs, 108–110
Skill set, 127

Skunk works, 11
Smith, C. Harold, 8
Software
    bug, discovery, 196
    design, 280
        skill requirement, 129
    engineers, usage, 86–87
    parallel development, 246
Source data, impact. *See* Financial data
Spreadsheet templates, usage, 220
Stage-Gate, 14
    development, 13
    process, 94
Standard development processes, 15–16
Step definitions, clarity, 165–166
Step level, project management, 166
Step-level resource assignments, 143
Step libraries, 286–287
Step-specific resource needs, 126
Step-specific team, self-coordination, 175
Strategy. *See* Product strategy
    information transformation, 24–28
Structured development process
    definition, 364
    limitations, 156
Structured product development process,
    173
Sub-assemblies, 25
Subcontractors, usage, 145
Subprojects, example, 247
Sub-systems, development, 244
Sub-teams, tasks, 257
Suppliers
    integration, 266–267
    partners, 266–267
Swat teams, 11
Systems-enabled collaborative development,
    265
System integration. *See* Development chain
    management system, integration

Tabulating Machine Company, 7
Tasks
    knowledge, linking, 287–289
    scheduling, 134
Teams, project. *See also* Project teams
    boundaries, 199

coordinating mechanisms, usage, 200
    leaders, planning, 64–65
    members, interaction, 252
    organization, 14
    resources, assignment, 208
Technology
    development projects, product develop-
        ment projects (integration), 247–248
    planning, 347–348
    roadmap, definition, 364
Technology-based companies, 23
Technology-driven companies, 339
Technology-intensive companies, 20
Templates
    research & development, 281–282
    usage, spreadsheet, 220
Tiger teams, 11
Time
    collection. *See* Closed-loop time collection
        system, 146
    contingency, assignment, 133
Time elements, 307
Time to market (TTM)
    decrease, 76
    definition, 364
    improvement, 69
    increase, 167
    objectives, 36
Time to market (TTM) Generation, 5, 29
    adoption curve, 19–20
    benefits, 17–18, 49, 155
    cross-functional project team organiza-
        tional model, 11
    improvement, 15, 32
    innovation, 14
    practices, 49
        adoption, 18–20
        improvement, 18
Time-phased planned technologies, 347
Time-phasing steps, resources, 126–127
Time-to-market generation, 12–20
    definition, 364
Time-to-market reduction, optimization, 183
Tolerances. *See* Project tolerances
Top-down architecture, 166
    bottom-up architecture, contrast, 166–167
Top-down/bottom-up process, integration, 74
Top-down program plan, 257

Top-down project planning, 15
Top-down resource planning, definition, 364
TTM. *See* Time to market

U.S. Army, 9
U.S. Department of War, 7
U.S. Navy, 9
Utilization. *See* Research & development, uti-
    lization; Research & development
    capacity
  baseline, 96
    establishment, 57–58
  importance, 31
  improvement, 117, 146
    reinvestment, 52
  management, 91
  performance expectations, 61
  reports, 91
  standards/goals, 60
  targets, 95

Vendors
  approach, customer/vendor, 10
  invoice, matching, 26
Vertical organization. *See* Approvals
Virtual project colocation environment, 188
Virtual situation room, usage, 201
Voicemail, usage, 185–186

Web-based collaboration workspaces,
    269
Web-based system, 345
Web-based workspace, usage, 269–270
Weighted ROI, 331
Wheelwright, Steve, 13
Work
  breakdown structure, 166
  estimates, roll-up, 155–156
Work management
  contrast, project planning, 163, 167–168
  definition, 364
  level, consistency, 168
Workload
  estimates, 73–74
    reconciliation, 137, 138
  management, 135
  reconciliation, 121, 134–136
  requirement, 124
  shifting, 133
Workplan review, 136
Worksheet organization, 123–124
World War I, 9
World War II, armaments (invention/produc-
    tion), 9
Wright, Orville, 7
Wright, Wilbur, 7

Year to date (YTD) measurement, 69